W9-CKT-050

Lost Woods

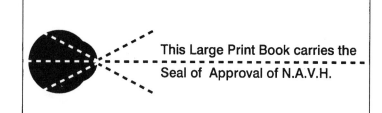

This Large Print Book carries the
Seal of Approval of N.A.V.H.

LOST WOODS

The Discovered Writing of Rachel Carson

**EDITED AND WITH
AN INTRODUCTION BY
LINDA LEAR**

Jefferson-Madison
Regional Library
Charlottesville, Virginia

Thorndike Press • Thorndike, Maine

15746330

Introduction and text (other than Carson's writing)
Copyright © 1998 by Linda Lear

All rights reserved.

Published in 1999 by arrangement with Beacon Press.

Thorndike Large Print ® Americana Series.

The tree indicium is a trademark of Thorndike Press.

The text of this Large Print edition is unabridged.
Other aspects of the book may vary from the original edition.

Set in 16 pt. Plantin by Rick Gundberg.

Printed in the United States on permanent paper.

Library of Congress Cataloging in Publication Data

Carson, Rachel, 1907–1964.
  [Selections. 1998]
  Lost woods : the discovered writing of Rachel Carson /
edited and with an introduction by Linda Lear.
    p.   cm.
  Originally published: Boston : Beacon Press, 1998.
  Includes bibliographical references.
  ISBN 0-7862-1697-2 (lg. print : hc : alk. paper)
  1. Nature. 2. Wildlife conservation. 3. Marine ecology.
  4. Bird watching. I. Lear, Linda J., 1940– II. Title.
QH81.C3546 1999
  570—dc21   98-44509

To my loving husband
John W. Nickum, Jr.

With gratitude
to Ruth Brinkmann Jerome
and
Ruth Jury Scott
whose lives have so richly blessed my own

Contents

Introduction

Rachel Carson's literary legacy is only four books. But those four books are enough to have changed how humankind regards the living world and the future of life on this earth. Her literary reputation rests primarily on two of them: *The Sea Around Us* (1951) and *Silent Spring* (1962), a book that changed the course of history.

The magnitude of Carson's impact on the public's understanding of such issues as ecology and environmental change still astonishes. Two volumes of her trilogy on the life of the sea, *The Sea Around Us* and *The Edge of the Sea*, were serialized in the *New Yorker*, and all three, including *Under the Sea-Wind* (1941), appeared on the *New York Times* best-seller list for months on end. *The Sea Around Us* maintained its place for a record eighty-six weeks and was eventually translated into more than forty languages. *Silent Spring* was also serialized in the *New Yorker*, making Carson the first woman writer to have three of her works introduced in its pages by 1962. It was translated into many languages, and still sells over 25,000 copies every year. Rachel

Carson had garnered an international reputation as a natural scientist and public voice for the care of the earth by the time of her death in 1964. She was the most acclaimed science writer of her generation and a literary figure of first rank.

The purpose of *Lost Woods*, this collection of Carson's undiscovered and little known writing, is to give the reader what is missing from the more famous body of Carson's work — a sense of her evolution as a natural scientist and a creative writer. Carson's unpublished and heretofore unknown literary output only heightens her importance as an environmental thinker. In this anthology, Carson's public and private voice speaks to our human condition and to the condition of our earth at the end of the millennium. Encompassing youthful writing, newspaper essays, field journals, speeches, articles, and letters, *Lost Woods* intimately reveals the intellectual process by which Carson became not just a literary celebrity, but one of the century's most important writers and social commentators, whose call to alarm took us all in a new direction and was the catalyst for the contemporary environmental movement.

The pace and pressure of Rachel Carson's life mitigated against there ever being a large body of writing. By nature she worked slowly and methodically, unwilling to move from one sentence to another until the first met her syntactical and lyrical satisfaction. She revised endlessly, read everything out loud, and then had it read back to her until she was satisfied with its tone, alliteration, and clarity. A perfectionist in form and structure, Carson was also a meticulous researcher whose demand for accuracy was legendary among her government colleagues, assistants, and editors.

It was gratifying for me to learn that Carson never finished a manuscript or an article on time with the exception of the feature stories she wrote for the Baltimore *Sun* in the 1930s. But it was heartrending to piece together the nearly overwhelming burden of family responsibility and emotional demand that prevented her from achieving the corpus of work that she dreamed of producing, and had the talent and vision to create.

Beginning in the late 1930s, Carson supported herself, her mother, her sister, and later her sister's two daughters, and her grand-nephew whom she adopted in 1957. A fifteen-year career in the federal government as an aquatic biologist and editor relegated her writing to evenings and time snatched between weekend obligations, yet also deepened her experience of the living world and her commitment to preserving it.

The literary success of *The Sea Around Us* brought a measure of financial security and enabled Carson to devote all her time to her writing after 1952. She enjoyed only a few years of freedom before her mother's physical decline, her niece's death, and the needs of a young child again stole her creative time and taxed her emotional stamina. The last five years of Carson's life were a race against time and the course of terminal illness. Fighting a misdiagnosed and aggressive breast cancer, Carson endured the side effects of treatment and the ravages of what she called a "catalogue of illnesses" to complete *Silent Spring* and to defend it. What is remarkable is not that Carson produced such a small body of work, but that she was able to produce it at all.

Rachel Carson had plans for at least four other major works. She had been collecting material for a scientific study of evolution, and had a book contract for another, more philosophical examination of ecology. She had started to revise and expand an earlier magazine article on exploring the natural world with children, and she was intrigued by the new discoveries in atmospheric science and climate and hoped to write something in this emerging field. Carson's literary papers display a full range of topics that she had, in one way or another, committed herself to writing about, and many more that she hoped one day to have the time to pursue. But time ran out in April 1964.

Lost Woods helps us fill the gap between Carson's wishes and her accomplishments. Selections from her field notebooks and especially her public speeches give Rachel Carson a voice for generations who neither heard her speak nor saw her on the few television appearances she made. Although she never thought of herself as a public figure, she became one, and was an accomplished public speaker as well, whose integrity captured the attention of the politically powerful and the average citizen alike. Her articles on the natural history of the Chesapeake region, her political acumen displayed in several editorial page letters, and her support of a wide variety of conservation and preservation efforts provide new facets to the better known lyrical writer on the sea, and trenchant critic of toxic chemicals.

The writings selected for *Lost Woods* are, for the most part, those I discovered in the course of my research for *Rachel Carson: Witness for Nature*, in her papers at the Beinecke Rare Book and Manuscript Library at Yale University. They have been chosen for their literary quality, as examples of important environmental thinking, and for the creative insight they provide into Carson's evolution as a scientist and science writer.

Several provide evidence of subjects about which Carson had an intense interest but never had opportunity to write about in any great depth. Other selections, including an article that appeared in *Holiday*, the preface to the second edition of *The Sea Around Us*, and the "Fable for Tomorrow" from *Silent Spring*, were published during Carson's lifetime and merit special attention. Some were published posthumously and are included here because of their scientific and literary quality and their biographical importance. A few others were published in journals, in newspapers, or as government documents and are no longer in print.

Carson once told her friend Dorothy Freeman that she considered her contributions to scientific fact less important than her attempts to awaken an emotional response to the world of nature. Describing the intent of her writing in a 1956 article on exploring nature with children, Carson wrote, "Once the emotions have been aroused — a sense of the beautiful, the excitement of the new and the unknown, a feeling of sympathy, pity, admiration or love — then we wish for knowledge about the object of our emotional response. Once found, it has lasting meaning."

The woman, the scientist, the reformer, and especially the writer who emerges from this collection elegantly combined science and emotion, reason and humanity. *Lost Woods* gives Rachel Carson a new and more complete voice for nature.

Here one will find further evidence of the centrality of ecological relationships in Carson's thought, and of her understanding of environmental ontology — the wholeness of nature. Displayed here as well is Carson's deep anxiety over the future of nuclear technology and how it might alter the intricate fabric of life.

But these selections also show for the first time that Carson's well-known support of wilderness preservation and wildlife conservation was particularly directed to the preservation of the nation's remaining wild seashores — areas that were fast disappearing in her own time.

Two selections testify to Carson's deep interest in animal rights. These issues were a natural extension of her lifelong reverence for life. Had she lived longer, she certainly would have become more politically active and would have written in support of humane treatment of laboratory and farm animals.

Carson had only begun to explore the evidence of global climate change before she died. Research for her 1957 television script on clouds revitalized her early interest in atmospheric temperature and wind which she had studied and written about in *The Sea Around Us*. She wanted to pursue this subject, and was convinced even then of the important relationship between human activity and alterations in climate.

Carson has much to teach deep ecologists and environmental policy makers as they wrestle with the moral dilemma of whether or not to moderate their demands and conform to political reality. Many will find insight in Carson's editorial writing and in her speeches on how to create conditions conducive to environmental reform. Carson's understanding of political process and the need for flexibility and compromise, as well as for intellectual toughness, can only encourage those who, like her, are searching for ways to move a democratic system off dead center.

Finally I have chosen to include several intimate pieces of writing. Carson's field notebooks are filled with incisive biological and ecological observation, but they also contain lyric vignettes that capture moments of dazzling humility. Similarly in letters to her friends and Maine neighbors Dorothy and Stanley Freeman, to Curtis and Nellie Lee Bok, and to her physician, George Crile, Jr., Carson reveals her deep love for the living world, and in the process, her quiet courage.

Not only does *Lost Woods* contain the discovered writing of one of the great writers and thinkers of our time, but it also illuminates a whole new Rachel Carson. Her outwardly calm and placid public life disguised the private passion as well as the complexity of her love for the natural world and her commitment to its wholeness. It is my hope that in this new collection of Rachel Carson's writing, readers may find and appreciate her varied and timeless voice.

Part One

The writings in Part One reflect the diversity of Carson's early interests and her efforts to find subject and style for her writing. It opens with the essay "Undersea," a characteristically searching and lyrical piece, which inaugurated her public literary career when it was published in the Atlantic Monthly *in 1937. Part One ends with selections from "Mattamuskeet," one of five* Conservation in Action *pamphlets Carson wrote and edited for the U.S. Fish and Wildlife Service. She was the second female professional hired by the agency, and during her fifteen-year federal career there, rose from aquatic biologist to editor-in-chief of all Service publications.*

"Mattamuskeet" reflects the confident writing of a mature scientist who knows her subject, her audience, and her public mission to inform. It also exhibits Carson's understanding of the intricate ecology of a wildlife habitat and her desire to communicate the

importance of these ecological relationships.

Between these two selections is a telling example of childhood writing and several journalistic features written for the Baltimore Sun *that demonstrate Carson's lifelong interest in the preservation of wildlife, her skeptical view of human interference in the natural world, and her passionate interest in birds. Two unpublished literary fragments from the 1940s testify to Carson's growing sophistication as a naturalist and nature writer. Taken together, these writings offer a window to Rachel Carson's early ecological consciousness and her evolution as a natural scientist.*

∽ **1** ∽

[1937]

Undersea

"Undersea" was originally titled "The World of Waters" and written as an introduction to a U.S. Bureau of Fisheries brochure in 1935. Carson's supervisor correctly assessed it as too lyric for a government report and encouraged her to submit it to the Atlantic Monthly, *where it was published by editor Edward Weeks. "Undersea" subsequently became the basis of Carson's first book,* Under the Sea-Wind *(1941), which remained her favorite piece of writing.*

The title "Undersea" was suggested by the Atlantic's editor who was impressed with Carson's illumination of science "in such a way as to fire the imagination of the layman." Its publication marked Carson's literary debut as a writer of critical merit.

Here Carson surveys both the ordinary and

23

fantastic creatures of the sea from the immediate perspective of an underwater eye, making the mystery and beauty of that world accessible to the nonscientific reader. "Undersea" introduces two of Carson's signature themes: the ancient and enduring ecology that dominates ocean life, and the material immortality that encompasses even the smallest organism. From these four remarkable pages in the Atlantic, *Carson later admitted, "everything else followed."*

Who has known the ocean? Neither you nor I, with our earth-bound senses, know the foam and surge of the tide that beats over the crab hiding under the seaweed of his tide-pool home; or the lilt of the long, slow swells of mid-ocean, where shoals of wandering fish prey and are preyed upon, and the dolphin breaks the waves to breathe the upper atmosphere. Nor can we know the vicissitudes of life on the ocean floor, where the sunlight, filtering through a hundred feet of water, makes but a fleeting, bluish twilight, in which dwell sponge and mollusk and starfish and coral, where swarms of diminutive fish twinkle through the dusk like a silver rain of meteors, and eels lie in wait among the rocks. Even less is it given to man to descend those six incomprehensible miles into the recesses of the abyss, where reign utter silence and unvarying cold and eternal night.

To sense this world of waters known to the creatures of the sea we must shed our human perceptions of length and breadth and time and place, and enter vicariously into a universe of all-pervading water. For to the sea's children nothing is so important as the fluidity of their world. It is water that they breathe; water that brings them food; water through which they see, by filtered sunshine from which first the red rays, then the greens, and finally the purples have been strained; water through which they sense vibrations equivalent to sound. And indeed it is nothing more or less than sea water, in all its varying conditions of temperature, saltiness, and pressure, that forms the invisible barriers that confine each marine type within a special zone of life — one to the shore line, another to some submarine chasm on the far slopes of the continental shelf, and yet another, perhaps, to an imperceptibly defined stratum at mid-depths of ocean.

There are comparatively few living things whose shifting pattern of life embraces both land and sea. Such are the creatures of the tide pools among the rocks and of the mud flats sloping away from dune and beach grass to the water's edge. Between low water and the flotsam and jetsam of the high-tide mark, land and sea wage a never-ending conflict for possession.

As on land the coming of night brings a change over the face of field and forest, sending some wild things into the safe retreat of their burrows and bringing others forth to prowl and forage, so at ebb tide the creatures of the waters largely disappear from sight, and in their place come marauders from the land to search the tide pools and to probe the sands for the silent, waiting fauna of the shore.

Twice between succeeding dawns, as the waters abandon pursuit of the beckoning moon and fall back, foot by foot, periwinkle and starfish and crab are cast upon the mercy of the sands. Every heap of brine-drenched seaweed, every pool forgotten by the retreating sea in recess of sand or rock, offers sanctuary from sun and biting sand.

In the tide pools, seas in miniature, sponges of the simpler kinds encrust the rocks, each hungrily drawing in through its myriad mouths the nutriment-laden water. Starfishes and sea anemones are common dwellers in such rock-girt pools. Shell-less cousins of the snail, the naked sea slugs are spots of brilliant rose and bronze, spreading arborescent gills to the waters, while the tube worms, architects of the tide pools, fashion their conical dwellings of sand grains, cemented one against another in glistening mosaic.

On the sands the clams burrow down in search of coolness and moisture, and oysters close their all-excluding shells and wait for the return of the water. Crabs crowd into damp rock caverns, where periwinkles cling to the walls. Colonies of gnome-like shrimps find refuge under dripping strands of brown, leathery weed heaped on the beach.

Hard upon the retreating sea press invaders from the land. Shore birds patter along the beach by day, and legions of the ghost crab shuffle across the damp sands by night. Chief, perhaps, among the plunderers is man, probing the soft mud flats and dipping his nets into the shallow waters.

At last comes a tentative ripple, then another, and finally the full, surging sweep of the incoming tide. The folk of the pools awake — clams stir in the mud. Barnacles open their shells and begin a rhythmic sifting of the waters. One by one, brilliant-hued flowers blossom in the shallow water as tube worms extend cautious tentacles.

The ocean is a place of paradoxes. It is the home of the great white shark, two-thousand-pound killer of the seas, and of the hundred-foot blue whale, the largest animal that ever lived. It is also the home of living things so small that your two hands might scoop up as many of them as there are stars in the Milky Way. And it is because of the flowering of astronomical numbers of these diminutive plants, known as diatoms, that the surface waters of the ocean are in reality boundless pastures. Every marine animal, from the smallest to the sharks and whales, is ultimately dependent for its food upon these microscopic entities of the vegetable life of the ocean. Within their fragile walls, the sea performs a vital alchemy that utilizes the sterile chemical elements dissolved in the water and welds them with the torch of sunlight into the stuff of life. Only through this little-understood synthesis of proteins, fats, and carbohydrates by myriad plant "producers" is the mineral wealth of the sea made available to the animal "consumers" that browse as they float with the currents. Drifting endlessly, midway between the sea of air above and the depths of the abyss below, these strange creatures and the marine inflorescence that sustains them are called "plankton" — the wanderers.

Many of the fishes, as well as the bottom-dwelling mollusks and worms and starfish, begin life as temporary members of this roving company, for the ocean cradles their young in its surface waters. The sea is not a solicitous foster mother. The delicate eggs and fragile larvae are buffeted by storms raging across the open ocean and preyed upon by diminutive monsters, the hungry glassworms and comb jellies of the plankton.

These ocean pastures are also the domain of vast shoals of adult fishes: herring, anchovy, menhaden, and mackerel, feeding upon the animals of the plankton and in their turn preyed upon; for here the dogfish hunt in packs, and the ravenous bluefish, like roving buccaneers, take their booty where they find it.

Dropping downward a scant hundred feet to the white sand beneath, an undersea traveler would discover a land where the noonday sun is swathed in twilight blues and purples, and where the blackness of midnight is eerily aglow with the cold phosphorescence of living things. Dwelling among the crepuscular shadows of the ocean floor are creatures whose terrestrial counterparts are drab and commonplace, but which are themselves invested with delicate beauty by the sea. Crystal cones form the shells of pteropods or winged snails that drift downward from the surface to these dim regions by day; and the translucent spires of lovely *Ianthina* are tinged with Tyrian purple.

Other creatures of the sea's bottom may be

fantastic rather than beautiful. Spine-studded urchins, like rotund hedgehogs of the sea, tumble over the sands, where mollusks lie with slightly opened shells, busily straining the water for débris. Life flows on monotonously for these passive sifters of the currents, who move little or not at all from year to year. Among the rock ledges, eels and cunners forage greedily, while the lobster feels his way with nimble wariness through the perpetual twilight.

Farther out on the continental shelf, the ocean floor is scarred with deep ravines, perhaps the valleys of drowned rivers, and dotted with undersea plateaus. Hosts of fish graze on these submerged islands, which are richly carpeted with sluggish or sessile forms of life. Chief among the ground fish are haddock, cods, flounders and their mightier relative, the halibut. From these and shallower waters man, the predator, exacts a yearly tribute of nearly thirty billion pounds of fish.

If the underwater traveler might continue to explore the ocean floor, he would traverse miles of level prairie lands; he would ascend the sloping sides of hills; and he would skirt deep and ragged crevasses yawning suddenly at his feet. Through the gathering darkness, he would come at last to the edge of the continental shelf. The ceiling of the ocean would lie a hundred fathoms above him, and his feet would rest upon the brink of a slope that drops precipitously another mile, and then descends more gently into an inky void that is the abyss.

What human mind can visualize conditions in the uttermost depths of the ocean? Increasing with every foot of depth, enormous pressures reach, three thousand fathoms down, the inconceivable magnitude of three tons to every square inch of surface. In these silent deeps a glacial cold prevails, a bleak iciness which never varies, summer or winter, years melting into centuries, and centuries into ages of geologic time. There, too, darkness reigns — the blackness of primeval night in which the ocean came into being, unbroken, through aeons of succeeding time, by the gray light of dawn.

It is easy to understand why early students of the ocean believed these regions were devoid of life, but strange creatures have now been dredged from the depths to bear mute and fragmentary testimony concerning life in the abyss.

The "monsters" of the deep sea are small, voracious fishes with gaping, tooth-studded jaws, some with sensitive feelers serving the function of eyes, others bearing luminous torches or lures to search out or entice their living prey. Through the night of the abyss, the flickering lights of these foragers move to and fro. Many of the sessile bottom dwellers glow with a strange radiance suffusing the entire body, while other swimming creatures may have tiny, glittering lights picked out in rows and patterns. The deep-sea prawn and the abyssal cuttlefish eject a luminous cloud, and under cover of this pillar of fire escape from their enemies.

Monotones of red and brown and lustreless black are the prevailing colors in the deep sea, allowing the wearers to reflect the minimum of the phosphorescent gleams, and to blend into the safe obscurity of the surrounding gloom.

On the muddy bottom of the abyss, treacherous oozes threaten to engulf small scavengers as they busily sift the débris for food. Crabs and prawns pick their way over the yielding mud on stilt-like legs; sea spiders creep over sponges raised on delicate stalks above the slime.

Because the last vestige of plant life was left behind in the shallow zone penetrated by the rays of the sun, the inhabitants of these depths contrast strangely with the self-supporting assemblage of the surface waters. Preying one upon another, the abyssal creatures are ultimately dependent upon the slow rain of dead plants and animals from above. Every living thing of the ocean, plant and animal alike, returns to the water at the end of its own life span the materials that had been temporarily assembled to form its body. So there descends into the depths a gentle, never-ending rain of the disintegrating particles of what once were living creatures of the sunlit surface waters, or of those twilight regions beneath.

Here in the sea mingle elements which, in their long and amazing history, have lent life and strength and beauty to a bewildering variety of living creatures. Ions of calcium, now free in the water, were borrowed years ago from the sea to form part of the protective armor of a mollusk, returned to the main reservoir when their temporary owner had ceased to have need of them, and later incorporated into the delicate statuary of a coral reef. Here are atoms of silica, once imprisoned in a layer of flint in subterranean darkness; later, within the fragile shell of a diatom, tossed by waves and warmed by the sun; and again entering into the exquisite structure of a radiolarian shell, that miracle of ephemeral beauty that might be the work of a fairy glass-blower with a snowflake as his pattern.

Except for precipitous slopes and regions swept bare by submarine currents, the ocean floor is covered with primeval oozes in which there have been accumulating for aeons deposits of varied origin; earth-born materials freighted seaward by rivers or worn from the shores of continents by the ceaseless grinding of waves; volcanic dust transported long distances by wind, floating lightly on the surface and eventually sinking into the depths to mingle with the products of no less mighty eruptions of submarine volcanoes; spherules of iron and nickel from interstellar space; and substances of organic origin — the silicious skeletons of Radiolaria and the frustules of diatoms, the limey remains of algae and corals, and the shells of minute Foraminifera and delicate pelagic snails.

While the bottoms near the shore are covered with detritus from the land, the remains of the floating and swimming creatures of the sea prevail in the deep waters of the open ocean. Beneath tropical seas, in depths of 1000 to 1500 fathoms, calcareous oozes cover nearly a third of the ocean floor; while the colder waters of the temperate and polar regions release to the underlying bottom the silicious remains of diatoms and Radiolaria. In the red clay that carpets the great deeps at 3000 fathoms or more, such delicate skeletons are extremely rare. Among the few organic remains not dissolved before they reach these cold and silent depths are the ear bones of whales and the teeth of sharks.

Thus we see the parts of the plan fall into place: the water receiving from earth and air the simple materials, storing them up until the gathering energy of the spring sun wakens the sleeping plants to a burst of dynamic activity, hungry swarms of planktonic animals growing and multiplying upon the abundant plants, and themselves falling prey to the shoals of fish; all, in the end, to be redissolved into their component substances when the inexorable laws of the sea demand it. Individual elements are lost to view, only to reappear again and again in different incarnations in a kind of material immortality. Kindred forces to those which, in some period inconceivably remote, gave birth to that primeval bit of protoplasm tossing on the ancient seas continue their mighty and incomprehensible work. Against this cosmic background the life span of a particular plant or animal appears, not as a drama complete in itself, but only as a brief interlude in a panorama of endless change.

ᷧ 2 ᷧ

[1922]

My Favorite Recreation

Rachel Carson knew from her earliest conscious memory that she wanted to be a writer. A solitary child, she read voraciously and was particularly influenced by the children's literary magazine St. Nicholas, *which not only offered writing of exceptional literary quality, but also awarded prizes for and published children's work. Carson submitted five stories in all and in doing so joined the company of such future literary luminaries as William Faulkner, F. Scott Fitzgerald, e.e. cummings, S. Eliot Morison, Edna St. Vincent Millay, and E. B. White, all of whom won prizes and were published in the pages of* St. Nicholas.

Carson's last story appeared when she was fifteen. She had already achieved the esteemed status of "Honor Member" of the St. Nicholas *League and been paid $10 for one of them. This* *story about exploring the Pennsylvania hills was her first about nature and was submitted in the category of "My Favorite Recreation." It shows something of Carson's already acute observation of the natural world and is noteworthy for the inclusion of her favorite bird, the wood thrush.*

The call of the trail on that dewy May morning was too strong to withstand. The sun was barely an hour high when Pal and I set off for a day of our favorite sport with a lunch-box, a canteen, a note-book, and a camera. Your experienced woodsman will say that we were going birds'-nesting — in the most approved fashion.

Soon our trail turned aside into deeper woodland. It wound up a gently sloping hill, carpeted with fragrant pine-needles. It was our own discovery, Pal's and mine, and the fact gave us a thrill of exultation. It was the sort of place that awes you by its majestic silence, interrupted only by the rustling breeze and the distant tinkle of water.

Near at hand we heard the cheery "witchery, witchery," of the Maryland yellow-throat. For half an hour we trailed him, until we came out on a sunny slope. There in some low bushes we found the nest, containing four jewel-like eggs. To the little owner's consternation, we came close enough to snap a picture.

Countless discoveries made the day memorable: the bob-white's nest, tightly packed with eggs, the oriole's äerial cradle, the frame-work of sticks which the cuckoo calls a nest, and the lichen-covered home of the humming-bird.

Late in the afternoon a penetrating "Teacher! *teacher!* TEACHER!" reached our ears. An oven-bird! A careful search revealed his nest, a little round ball of grass, securely hidden on the ground.

The cool of approaching night settled. The wood-thrushes trilled their golden melody. The setting sun transformed the sky into a sea of blue and gold. A vesper-sparrow sang his evening lullaby. We turned slowly homeward, gloriously tired, gloriously happy!

∿ 3 ∿

[1938]

Fight for Wildlife Pushes Ahead/
Chesapeake Eels Seek the Sargasso Sea

Carson completed her M.A. in Zoology at Johns Hopkins University in Baltimore, Maryland, in 1932. The Depression dashed her hopes of going on for a doctorate but she continued to teach part-time at the University of Maryland while she tried to find a college position. Although she thought she had forever abandoned a writing career, economic necessity, compounded by the death of her father in 1935 and her assumption of the role of head of household, forced her to return to writing.

Research for radio scripts she was writing for the Bureau of Fisheries served as the basis of feature articles on Maryland's natural history which she sent to the local newspaper, the Baltimore Sun. *Sunday editor Mark Watson was impressed with Carson's lucid style and her* scientific accuracy and published as many of her articles as he could, sometimes sending those he could not use to affiliated newspapers.

Much of Carson's newspaper writing concerns the population and habitat changes of mid-Atlantic fish and wildlife and reflects the research of a thoroughly competent marine biologist. They show Carson's already broad interest in the conservation of resources, her special interest in wildlife, her concern with the impact of human exploitation on wildlife habitats, and her fascination with the intricate processes of nature.

Carson's interest in eels began during her summer study at the Marine Biological Laboratory at Woods Hole, Massachusetts, where she saw the ocean for the first time in the summer of 1929. Later, in a protozoology experiment at Johns Hopkins, she observed the effect changes in the salinity of sea water had on the behavior of the eel. Carson's fascination with these beautiful and unfathomable creatures appears most significantly in Book III of Under the Sea-Wind, where the central character is Anguilla, a European eel who ends her maturation with a two-hundred-mile journey to the open sea.

Carson's brief but successful journalistic career with the Baltimore Sun was an important apprenticeship in writing science for the public. It established her identity as a writer who had discovered what she wanted to write about.

Fight for Wildlife Pushes Ahead

The inescapable fact that the decline of wildlife is linked with human destinies is being driven home by conservation the nation over. Wildlife, it is pointed out, is dwindling because its home is being destroyed. But the home of wildlife is also our home.

One of the most startling pictures painted by those who are fighting for conservation of natural resources is that of the speed with which the work of destruction has been accomplished. It is not necessary to go back to Colonial times for contrast. A scant hundred years ago, more than half of America was unspoiled wilderness. Wild swans, geese and brant were to be found in every marsh and slough and prairie pot-hole; slaves in the Chesapeake Bay country were fed on canvasbacks until they are said to have revolted at the fare; wild turkey, grouse and other upland game birds were incredibly abundant. Antelope ranged the Western plains in numbers perhaps equalling the bison, and from coast to coast the bugling of the elk resounded in the forests.

A hundred years ago, Audubon, the artist naturalist, saw from his Kentucky village the skies literally clouded with the flocks of the passenger pigeon. He estimated that more than a billion birds must have passed over-head in a four-day flight. When the beechnuts were ripe, the pigeons flew 200 miles in a day to feed on them, and forest areas of more than a hundred square miles were so densely packed with roosting birds that the trees broke under their weight.

A hundred years ago, salmon still ran in the rivers of New England wherever dams had not blocked their passage and mills poisoned their spawning beds. The spring run of the alewife, or river herring, was an important event of the year to villagers on New England rivers, and shad poured into the Susquehanna, the Delaware and other coastal streams in such numbers that the shallow waters foamed with their passage. Sturgeon leaped in the waters of the Great Lakes, where the sails of the first fishing vessels in those inland seas moved over Erie, Huron and Michigan.

A hundred years ago the flights of migratory waterfowl winging their way southward along the Mississippi flyway passed along the dividing line between the known and the unknown halves of the continent. To the west, beyond miles of prairie still unbroken by the plow, the sun set over untamed Rockies; eastward, a sprinkling of farms and villages traced out the Ohio and the Tennessee and the wall of the Appalachians hid the lights of the Seaboard cities, where, alone on all the continent, were dense settlements.

But what of wildlife today? Government service, whose business it is to know conditions, paint a general picture of scarcity and depletion. The last heath hen perished on the island of Martha's Vineyard in 1933, and the passenger pigeon is now a creature of legend. Salmon are virtually gone from the rivers of New England, and the Atlantic Coast shad fisheries have declined some 80 per cent within half a century. Waterfowl flights fell in 1933 and 1934, and although Government regulations plus the establishment of sanctuaries have resulted in some improvement, the plight of certain species, notably canvasback and redhead duck, remains serious. The ranks of elk were so thinned by 1904 that domestication was urged as the only means of preventing their extinction. Although prong-horn antelope are now on the increase within refuges and reservations, they are reduced from some 30,000,000 or 40,000,000 to about 60,000. Mountain goats, moose and grizzly bear are also on the wane.*

* Ed.: The population of all big game animals today is much larger than in 1938. Pronghorn elk, mountain goats, and moose are now hunted legally. Grizzley bears have rebounded with management to such a degree that they are a problem in some areas.

Yet this mere remnant of wildlife supports a resource estimated by business interests as worth considerably more than a billion dollars a year in cash turnover. Sportsmen's expenditures run to three-quarters of a billion annually, while expenditures of others in the enjoyment of wildlife are estimated at something over half a billion. Every year, more than 5,000,000 automobiles carry sportsmen to hunting and fishing grounds, the mileage used being equivalent to the actual consumption of 87,000 automobiles. In the States of New York and New Jersey, about 2,000 boats are licensed to carry fishing parties, and the fish and game commissioners of the latter State estimate that each season more than a million salt-water anglers are attracted to its shores.

Such figures carry convincing proof that preservation of wildlife is good business. However, the job of conservation that is being urged this week has a deeper significance than the restoration of wildlife alone.* For three cen-

---

Canvasback and redhead duck populations bounced back in the 1950s, but then declined. They have remained steady in recent years, although at levels that are low even when compared to what they were in 1938.

* Ed.: The week in which this article was published had been proclaimed National Wildlife Restoration Week.

turies we have been busy upsetting the balance of nature by draining marshland, cutting timber, plowing under the grasses that carpeted the prairies. Drainage operations, intended to reclaim more land for agriculture, have directly affected millions of acres of waterfowl nesting grounds, and indirectly destroyed additional millions by lowering the water tables of the soil from 10 to 60 feet within a score of years.

The story of lower Klamath Lake in Oregon, once described by Theodore Roosevelt as "one of the greatest wild-fowl nurseries in the United States," has been repeated many times in the case of other areas drained in the name of progress. Klamath Lake was drained at considerable expense to convert the region to agricultural use, later devastated by numerous fires, and finally abandoned because it was found impossible to sweeten the soil of these former marshlands for agricultural crops. There is now talk of reflooding it!

But as long as it was only the ducks and their kind that were being pushed nearer and nearer the brink of extinction, the cause of wildlife had few champions. Then one day — less than four years ago — the winds blowing over the Western prairies picked up the soil that had no anchor because the grass was gone and carried it eastward. People in Pennsylvania looked up to see the sky darkened by dust from the fields of Kansas, and New York farmers received a donation of soil from Nebraska. The words "dust bowl" and "resettlement" became part of our national vocabulary.

The program of national conservation agencies that is being put before America this week is no mere sentimental plea for birds and fish and big-game animals. It means checking the spread of the dust bowl, and perhaps in time binding its swirling sands once more with the tough roots of prairie grasses. It means reforestation of hillsides so that the melting snows may be held in the ground that is dying of thirst. It means giving back to the waterfowl and the muskrat a few million acres of land which nature meant to be marsh forever. [. . .]

Chesapeake Eels Seek the Sargasso Sea

From every river and stream along the whole Atlantic Coast, eels are hurrying to the sea. Reaching salt water, they will strike out south and east to the Sargasso, there to mingle with other eel hordes which have made the longer westward crossing from Europe. From Greenland, Labrador, the United States,

Mexico, and the West Indies; from Scandinavia, Germany, Belgium, France and the British Isles, eels go at spawning time to those mid-oceanic meadows of brown sargassum weed.

So the most remarkable of all Chesapeake Bay fishes is born in alien waters. Before it is half as long or as thick as a man's thumb it makes a journey across 1,000 miles of strange, wild waters without benefit of chart or compass, finding the shores from which its parents came a year and a half before. In bays, rivers and streams it feeds and grows for ten years, perhaps fifteen or twenty. At last, obeying an instinct as old as the tribe of eels, it sets out on the return journey to the Sargasso to produce its young and itself to die. Thus is the life cycle of the eel completed.

Some 2,000 years ago Aristotle declared that eels were generated spontaneously from mud. Even today there are people who still subscribe to the ancient belief that a horse hair falling into water becomes an eel. Within the past two decades, reputable scientists knew little more than the fact that spring and fall the eels are running in the rivers — in the fall the old eels are bound for the sea, in the spring the young are ascending every bay and river estuary.

Four and a quarter centuries after the Nina,

the Pinta and the Santa Maria crossed the dread sea of floating sargassum weed another explorer, Danish Johannes Schmidt, sailed over a spot in the Sargasso, south of Bermuda and a thousand miles east of Florida, and declared it to be the breeding place of the eel. In twenty years of painstaking research he had literally strained the surface waters of the ocean for eel larvae, finding younger and younger stages all the way across the Atlantic from Europe, until at last he found the youngest stage of all and knew he had reached the birthplace of the eels.

Let's picture the spawning journey of a Chesapeake Bay eel. Part of it we know from observed fact; part we shall have to supply from imagination aided by our knowledge of later happenings. If our eel lives far up toward the headwaters of one of the rivers tributary to the bay, it is almost certain to be a female, for the males usually remain in salty or brackish water near the river mouths.

Other autumns to the number of ten, fifteen or twenty have come and gone, but our eel has never before felt the desire to leave the familiar mud banks dotted with crayfish burrows, the marshy banks where small fowl or water rats could now and then be seized, the forests of water weeds where hunting was good for minnows, sunfish, and perch.

Now physical maturity has attuned her to the call of seaward hurrying water. One dark night, when wind ruffles the surface of the river and clouds hide the moon, she slips away downstream on the journey which she will never retrace. Hiding by day, drifting with the currents by night, she finds the river ever widening, the channels deepening, the water bringing unfamiliar tastes to her keen senses.

She is not alone; more and more eels have joined the caravan. Probably as their numbers increase and the strange, bitter tang of salt grows stronger in the water the excitement of the eels grows, they travel faster, rest less often. In the lower estuary of the river the male eels have been living, growing fat on shellfish, worms and water plants — on shad and herring looted from fishermen's gill nets in the spring. Compared with the 3- to 4-foot females, however, the males are dwarfs, never growing to a greater length than two feet.

Gradually the river garb of olive brown is changed for a coat of glistening black with under parts of silver: This is the dress worn only by eels about to undertake the far journey to the Sargasso. The snouts become high and compressed, probably owing to some sharpening of the sense of smell; the eyes become twice their former size, as though in preparation for the descent along darkening sea lanes.

After the eels leave our shores nothing more is seen of them. The only clue to their destination is the finding of the newly hatched larvae floating nearly a thousand feet below the surface of the Sargasso Sea. How do the migrants find their way? Perhaps one man's guess is as good as another. The English naturalist Henry Williamson suggests the eels find the Gulf Stream and swim against its current, their keen nostrils scenting in its warm waters the rotting sargassum weed.*

Even more puzzling, how do the fragile larvae, as transparent as glass and flattened like a willow leaf, find their way back to the shores from which the parents came? And how do the children of the American and European eels return to the proper continent?

When a few months old and less than an inch long, eel larvae begin their homeward mi-

* Ed.: The mysterious migration of the eels to the Sargasso to spawn remains one of the great riddles of zoology. Larvae have been found drifting in the ocean currents in both eastern and western directions from the Sargasso, but no mature eels have been caught in the open ocean. It is simply not known by what mechanism the adult eel manages to find its way out to the Sargasso from the freshwater estuaries in which they spend their adult lives.

gration, aided by the movements of the water currents. Since the breeding grounds of the European and American eels overlap, the larvae of the two species travel together for a time. (The European eel has a larger number of vertebrae and so may be distinguished even as a larva.) Finally, the two great stream of larvae begin to diverge, the American eels turning westward, the European eastward.

From January to March, when something less than a year old, American eel babies are arriving in coastal waters off the Chesapeake, off New England somewhat later. At that time European eel babies are somewhere in mid-Atlantic and will not reach shore until they are 3 years old.

As a partial explanation of the infallible homing instinct of the two species of eels, scientists point out that the American eel undergoes a change from the flat, leaflike larva to the rounded "glass-eel" stage when it is only a year old, while the European eel requires two years more. Until this stage is reached, scientists say, the young eels feel no urge to seek the coast — therefore there is no chance that a young European eel will make port in the wrong continent.

When the young eels begin to enter our rivers they are from 2 to 3 ½ inches long and practically colorless except for the eyes. The segregation of the sexes is already apparent, the males remaining in tidal marshes and brackish river mouths, the females pushing upstream, clambering over falls, up dams, even over damp rocks. "Elvers," as they are called, swim at or near the surface, often forming an unbroken procession extending for miles along the edges of rivers or creeks.

In some European rivers the elvers themselves are taken for food; in others they are caught for stocking in other rivers less plentifully supplied with eels for the ready European markets.

Although not in great favor in local markets, eels support one of the more important Chesapeake fisheries. Out of the thirty-six varieties of fishes produced in Maryland waters, eels rank ninth in poundage, eighth in value. About a quarter of a million pounds are produced in Maryland, this figure being a little more than half of the total Chesapeake yield. Most of the output is shipped to New York and other distant markets, but part of the catch is consumed locally and part is used as bait, especially on trotlines fished for hard crabs. The Maryland Legislature once (about 1890) spent $3,400 in an attempt to exterminate eels because of their frequent raids on fishes caught in gill nets.

[1944]

Ace of Nature's Aviators

*It was typical of Rachel Carson's literary crafts-
manship and scientific understanding that she
could find in the most mundane parts of the natural world some aspect that endowed the familiar
with a unique sense of worth, even redemptive
value. An early example was a feature article re-
habilitating the common starling which first appeared in the* Baltimore Sun *in early 1939.
Carson sold a revised version to* Nature Magazine *later that same year as "How About Citi-
zenship Papers for the Starling?" It brought
favorable comment from readers who were fascinated to learn something of the worthiness of this
much maligned bird.*

*Research that came across Carson's Fish and
Wildlife desk during the war years reinforced her
determination to write about scientific topics that*

would inform the public as well as make the hidden processes of nature understandable to the general reader. "Ace of Nature's Aviators" reports the discovery of the remarkable migration patterns of the "remote and mysterious" chimney swift. It began as a Department of Interior press release. Having revised it as a feature article, Carson offered it to the Baltimore Sun *and to* Reader's Digest, *but since she was in need of immediate money following an appendectomy, she hastily sold a condensed version to* Coronet *which was published as "Sky Dwellers" in November, 1945.*

If aviation engineers could apply the wisdom of the chimney swift, several troublesome problems of aeronautics could be solved. Pilots, for example, would never have to worry about the amount of gasoline in their tanks. The chimney swift refuels on the wing, spends almost its entire waking life in the air, and never, except by accident, touches the earth.

In creating one of her most efficient mechanisms for flight, Nature has fashioned the swift as a flying insect trap. Its beak is short, its mouth one of the widest in birddom. Its torpedo-shaped body and long, slender wings are built for speed and adapted to sudden twists and turns. From dawn till dusk the swift speeds open mouthed through the sky, straining insects out of the air. Although the bird's aerial existence involves a high rate of fuel consumption, its energy is perpetually renewed by the almost continuous intake of food.

Not only does it eat in the air, the chimney swift drinks and bathes on the wing, dipping to the surface of a pond for a momentary contact with the water; its courtship is aerial; it sometimes even dies in the sky. Probably it is less aware of the earth and its creatures than any other bird in the world. It never perches on a tree, never alights on the ground. Its whole existence is divided between the sky and a nocturnal resting place inside a chimney or a hollow tree.

For its mastery of the air, the swift has paid a strange penalty. Its feet have degenerated into little more than hooks, useless for perching or hopping as other birds do, but perfect for clinging to the wall of a chimney. As a result of its inability to perch, the swift's idea of going to bed is merely to hang itself up for the night against some vertical surface, its toes securely hooked in a crack or over a convenient projection. Its stubby tail, edged with a row of bristles, provides a useful prop.

The chimney swift is one of the few birds

unharmed by the white man's invasion of North America — it has actually profited by it. Ancestors of the modern chimney swift lived in great hollow trees. When pioneering Americans began to cut down the forests and to build cabins and houses, then churches, schools, and factories, the swifts discovered that a chimney is a first-rate substitute for a hollow tree. Almost to a bird they changed their habits.

In more isolated parts of the country, a few chimney swifts cling to old-fashioned ideas: they still nest in hollow trees. The western cousin of the chimney swift — Vaux's swift — only of recent years has begun to make the transition from trees to chimneys. So broadminded and so adaptable is the true chimney swift, however, that it has made the most of various other conveniences of civilization, nesting in abandoned buildings, in wells and cisterns, and in silos.

For its intra-chimney architecture the swift is equipped with enormously developed salivary glands. These secrete a thick, gluey saliva useful in fastening twigs together and in cementing the hammocklike nest to the wall of the chimney. The Chinese swift dispenses with twigs, fashions its entire nest of saliva, and so creates the principal ingredient of the delicacy known as bird's nest soup.

During the nesting season the salivary glands enlarge, providing copious supplies of the needed cement. Later they shrink, but the hollow spaces left in the cheeks of the swift are put to good use — the bird crams them full of insects to bring back to its hungry babies.

Nest building takes two to three weeks, even longer if the days are rainy and the glue melts. Every twig used is collected by an amazing method: the bird snatches them on the wing from trees and shrubbery. To this day ornithologists cannot agree whether it uses feet or beak in the process.

Swifts are devoted parents. The male and female take turns incubating the eggs during the nearly three weeks required for the young to hatch. Thereafter, both birds assume the chore of keeping the infant mouths filled with insects, a task that must be performed faithfully for about four weeks before the young swifts are able to take to the sky in their own behalf.

Unexplained as yet is the fact that observers have sometimes seen three adult birds tending a nest. The polite but wholly tentative theory is that the parents have engaged a "nursemaid." More realistic persons scoff at that and say the swift is polygamous. What the truth is, no one actually knows.

The chimney swift is rated the fastest small

bird in North America and has few natural enemies it needs to fear. Flight records for an Asiatic swift indicate speeds up to 200 miles an hour. It is debatable whether even a duck hawk can overtake one in straight-away flight. An occasional swift, however, may be snatched by a hawk as the birds are circling above a chimney, preparatory to entering for the night.

From one enemy — rain — the bird has no defense. Cold rains, long continued, wash the skies clear of insects. Deprived of food, the swifts weaken and die in great numbers. During one unseasonably wet June reports of dead swifts came from all over southern New England, wheelbarrow loads were taken from the chimneys of large mills, bushels from the base of a chimney at Clark University. Fortunately, such occurrences are comparatively rare.

The life story of the chimney swift has been pieced together by naturalists and amateur birders only with the greatest patience and perseverance. A bird who never perches on a tree where you can focus your binoculars on him, who never visits feeding stations, who spends almost all the daylight hours far above your head, and who, in the fall, vanishes so suddenly and completely that only within the past year has his winter home been discovered

— such a bird is not an easy subject for his would-be biographers.

Perhaps because of the very difficulty of the job, an extraordinary number of people seem to have interested themselves in the chimney swift, and have gone to endless trouble to learn its habits. One woman in Iowa had an imitation chimney, equipped with observation tower, built in her back yard so she could study the home life of the swifts that later nested in it. A West Virginia farmer suspended tin coffee cans in his chimney as an invitation to the swifts to nest there. The birds accepted, later permitted the farmer to raise the cans to the top of the chimney at intervals to photograph the young. The ornithologist and artist George Mitsch Sutton, as a young man in West Virginia, repeatedly climbed a tall church chimney and hung for hours, cramped, shivering, and wretched, just below the mouth of the chimney so he could make accurate notes on the wing movements of the swifts as they dropped into the chimney. And all over eastern United States and Canada, people have industriously banded chimney swifts in an effort to trace their migrations, until the total of swifts marked with identifying metal bands now exceeds 375,000.

Banding 375,000 chimney swifts is not as simple as it may seem. If you are one who insists upon lying abed in the chilly dawn or whose experience as a steeple jack is limited, don't take up swift banding as a hobby. It requires a certain Spartan fortitude. Unlike most birds, the insect-eating swifts cannot be attracted to cage traps baited with grain or other foods. They have to be caught as they emerge at daybreak from the large chimneys where thousands of them sleep, especially near the time of the fall migration.

Swift banders stand about on roofs, shivering as they wait for daylight and the birds. They risk their necks clambering to the tops of tall chimneys to set their traps. They incur the suspicion of the law as they skulk about empty buildings in the small hours. Despite these hazards, Constance and E. A. Everett of Minnesota once wrote cheerfully to an ornithological journal about "the fun of banding chimney swifts."

A few months ago the banders had their reward. Although many marked swifts had been recaptured, all recoveries had been made within the known summer range of the bird, and the winter home of the swift was still undiscovered. Then to the U.S. headquarters for the study of bird migration, the Fish and Wildlife Service, came a long, official envelope from the American Embassy at Lima, Peru. It contained thirteen bands, taken from chimney swifts shot by Indians in the jungles of Peru during the northern winter. Records showed that the birds had been banded in Tennessee, Illinois, Connecticut, Alabama, Georgia, and Ontario, dates of banding ranging from 1936 to 1940.

For the thirteen small birds, death won ornithological fame. The swift was the last North American bird whose winter range was unknown. The thirteen had now provided the solution of a major mystery of bird migration, had filled in the missing paragraphs in the biography of their race.

∽ 5 ∾

[1945]

Road of the Hawks

Rachel Carson's lifelong fascination with the sea was matched by an equally intense interest in birds, nurtured first in the company of her mother in the hills of western Pennsylvania. It was a passion she maintained throughout her life.

At her U.S. Fish and Wildlife job in wartime Washington, Carson's ornithological interests found outlet in company with other members of the newly organized Audubon Society of the District of Columbia. Carson was soon elected to the Society's Board of Directors where she worked with artist Roger Tory Peterson and other notable scientists. Society activities also provided the rare occasion for social gatherings and outings that Carson enjoyed.

One of the most popular Society trips was to the Hawk Mountain Sanctuary in eastern

Pennsylvania to watch the fall migration. In October, 1945, Carson, along with her Fish and Wildlife colleague and friend Shirley Briggs, who was an equally avid amateur ornithologist, spent two days at Hawk Mountain. Perched on a rocky promontory, braving a bone-chilling wind, Carson watched the hawks and took notes on their behavior. The following fragment from those field notes reveals how deeply Carson was affected by the spectacle of the hawks, but also shows how she related even non-marine experiences in nature to the ocean and to the ancient history of the earth.

They came by like brown leaves drifting on the wind. Sometimes a lone bird rode the air currents; sometimes several at a time, sweeping upward until they were only specks against the clouds or dropping down again toward the valley floor below us; sometimes a great burst of them milling and tossing, like the flurry of leaves when a sudden gust of wind shakes loose a new batch from the forest trees. [. . .] On the horizon to the north, formed by a series of seven peaks running almost at right angles to the ridge on which we sit, an indistinct blur takes form against the sky. Second by second the outlines sharpen. Soon the unmistakable silhouette of a hawk is etched on the gray. It is too soon to make out the identifying lines of wing and tail that mark him for one species or another. On he comes, following the left side of the ridge, high up. Sometimes he banks steeply and his outlines melt into the sky. Then a swift wing beat or two and we have him in our glasses again [. . .]

Now follows a long wait with no more hawks. I settle back against the rock behind me, seeking shelter from the wind, trying vainly to draw some physical comfort from the hard angularity of stone. The cold is bitter. The morning had seemed reasonably mild down in the valley, as we had our quick cups of coffee in the predawn blackness. But here on the mountain top we are in the sweep of all the winds out of a great emptiness of sky, and the cold seeps through to the very marrow of my bones. But cold, windy weather is hawk weather, and so I am glad, although I shiver and my nose reddens, and I look speculatively at my thermos of hot coffee. But that must last the day, and now it is only ten o'clock.

Mists are drifting over the valley. A grayness overhangs all the sky and the clouds seem heavy with unshed rain. It is an elemental landscape — a great rockpile atop a mountain, nearby a few trees that have been stripped and twisted by the mountain winds, a vast, pale, arching sky.

Perhaps it is not strange that I, who greatly love the sea, should find much in the mountains to remind me of it. I cannot watch the headlong descent of the hill streams without remembering that, though their journey be long, its end is in the sea. And always in these Appalachian highlands there are reminders of those ancient seas that more than once lay over all this land. Halfway up the steep path to the lookout is a cliff formed of sandstone; long ago it was laid down under shallow marine waters where strange and unfamiliar fishes swam; then the seas receded, the mountains were uplifted, and now wind and rain are crumbling the cliff away to the sandy particles that first composed it. And these whitened limestone rocks on which I am sitting — these, too, were formed under that Paleozoic ocean, of the myriad tiny skeletons of creatures that drifted in its water. Now I lie back with half closed eyes and try to realize that I am at the bottom of another ocean — an ocean of air on which the hawks are sailing.

∿ 6 ∿

[1946]

An Island I Remember

In the decade following the publication of Under the Sea-Wind *in 1941, Carson worked on an extended profile of the ocean which would become* The Sea Around Us. *In 1946, after almost ten years in the federal government, Carson had accumulated enough annual leave for a month's vacation in Boothbay Harbor, Maine, the site of a Bureau of Fisheries laboratory engaged in research on lobster reproduction, and a coastal area she had longed to visit.*

Carson along with her mother and her two cats rented a tiny, secluded pink cottage on the shore of the Sheepscot River west of town. It looked out to Indiantown Island, a mysterious, spruce-covered strip of lush forest where the wind rustled through the trees and at sunset a hermit thrush sang its eerie song.

Carson fell in love with the beauty of Maine that summer and determined that one day she would have a place of her own there. She told her friend Shirley Briggs in a letter, "the only reason I will ever come back [to Maryland] is that I don't have the brains enough to figure out a way to stay here the rest of my life." Seven years later the success of The Sea Around Us *enabled her to buy land and build a cottage along the Sheepscot on Southport Island.*

Among the unpublished fragments in Carson's papers, there is no more arresting example of Carson's ability to absorb her surroundings with all her senses, or of her pleasure in the diverse fabric of the natural world.

It was only a small island, perhaps a mile long and half as wide. The face it presented to the mainland shore was a dark wall of coniferous forest rising in solid, impenetrable blackness to where the tops of the spruces feathered out into a serrate line against the sky. There was no break in that wall anywhere that I could see, no suggestion of paths worn through island forests, no invitation to enter. At high tide the sea came up almost to the trees, with only little patches of light colored rock showing at the water line, like white daubs made by a painter's brush. As the tide ebbed and the water dropped lower on the rocks, the white patches grew and merged with each other, exposing the solid granite foundations of the island, so that now there was a high rocky rampart, on which grew the living green wall of the forest.

There was perhaps a quarter of a mile of water between the island and the mainland shore where our cabin stood, its screened porch at the sea wall and its back against a steep hillside where ferns were dark among the rocks and the branches of great hemlocks reached down to touch the ground. Day after day the island lay under the summer sun, with no sound coming from it, and nothing moving at the visible edges of the forest. On every low tide I could see a solemn line of cormorants standing on a rocky ledge that ran out from the south end of the island, their long necks extended skyward. The gulls were less broodingly protective of the approaches to the island forest, their presence about the shores of the island more casual as they perched on the weed-covered rocks while waiting for the turn of the tide.

About sundown the island, that had lain so silent all day long, began to come to life. Then the forms of large, dark birds could be seen moving among its trees, and hoarse cries that brought to mind thoughts of ancient, reptilian monsters came across the water. Sometimes one of the birds would emerge from the shadows and fly across to our shore, then revealing itself as a great blue heron out for an evening's fishing.

It was during those early evening hours that the sense of mystery that invested the island drew somehow closer about it, so that I wished even more to know what lay beyond the wall of dark spruces. Was there somewhere within it an open glade that held the sunlight? Or was there only solid forest from shore to shore? Perhaps it was all forest, for the island voice that came to us most clearly and beautifully each evening was the voice of a forest spirit, the hermit thrush. At the hour of the evening's beginning its broken, silvery cadences drifted with infinite deliberation across the water. Its phrases were filled with a beauty and a meaning that were not wholly of the present, as though the thrush were singing of other sunsets, extending far back beyond his personal memory, through eons of time when his forebears had known this place, and from spruce trees long since returned to earth had sung the beauty of the evening.

It was in the evenings, too, that I came to know the herring gulls as I had never known them before. The harbor gull — the gull of the fish wharves — is an opportunist. He sits with his fellows on the roofs above the harbor, or each on his wharf piling, and waits, knowing at what hour the refuse will be discharged from the fish house, or when the first of the returning fishing boats will appear on the horizon, to be met with excited cries. But the gulls of the island were different. They were fishermen, and like the men who handle the nets, they lived by their own toil.

I suppose a certain amount of regular fishing went on during the day, but I was especially aware of the excitement that attended the runs of young herring that came into our cove each evening.

It is strange to reflect on that twilight migration of the young herring that by day have moved widely through the coastal waters, but now are drawn, as the water darkens to black and silver, to follow the channels between the rocky foundations of the islands.

We would know the herring were coming by watching the behavior of the gulls. Most of the late afternoon they would have dozed on their rocky perches along the shore of the island. But as sunset neared and the shadows of the spruces began to build dark spires in the water, a stir of excitement would pass among the gulls. There would be a good deal of flying up and down the channel, as though scouts were coming and going. It seemed that some intelligence of the movements of the fish was being spread among the birds. More and more of the gulls would join in the scouting parties, until the whole flock was in movement, their sharp, staccato cries coming across the water.

When the water was glassy calm, holding the colors of the evening sky on its surface, we could time as exactly as the gulls the arrival of the herring in our cove. Suddenly the silken sheet would be dimpled by a thousand little noses pushing against the water film. It would be streaked by a thousand little ripples moving eagerly toward the shore. It would be shot through by a thousand silver needles as the fish, swimming just beneath the surface, disturbed the placid sheet. Then the herring would begin flipping into the air. It seemed it was always out of the corner of your eye that you saw them, and you never quite knew where to look for the next little herring skipping recklessly into the air in a sort of back somersault. They did it as though it were great fun — this rash defying of a strange and hostile element, the air. I believe it was a sort of play indulged in by these young children of the herring. They looked like silvery coins skipped along the surface. I never actually saw any of the youngsters caught in the air by a gull, but the quick eyes of the birds must certainly have been attracted by the bright flashes.

The gulls would greet the arrival of the herring schools with a frenzy of excitement, swooping, plunging, crying loudly. A gull does not dive as a tern does; he swoops and, not quite alighting, plucks his fish from the water. It takes a good eye and good timing. It is less graceful than the beautiful, clean dive of a tern, but perhaps it requires equal skill.

A night I especially remember there had been a large run of herring into the cove and it had come somewhat later than usual. The gulls, apparently determined to make their catch despite the gathering darkness, fished on until it was hard to understand how they could possibly see the fish. We could see their moving forms against the island — white, mothlike figures against the dark backdrop of the island forest, fluttering to and fro and all the while uttering their cries, in a scene out of some weird shadow world.

On sunny days the gulls would go aloft to ride the warm, ascending air currents. Up and up, sailing around in slow, wide circles, until they were almost lost to sight. I used to lie on my back on the dock, relaxing in the warm sunshine, and watch the gulls above me in the blue sky. Some were so high they were only white stars wheeling slowly in orbits of their own making.

It was possible to do a good deal of birding by ear alone, lying there on the dock half asleep. Once the sound had been identified by squinting through half-opened eyes, I knew without looking that the mouselike rustling and patter of very small feet on the dock, skirting my head and passing just beyond my outstretched arm, was the song sparrow on whose territory we were living. I knew that the soft "whuff, whuff" overhead was the wing beat of a gull, the bird passing so close that I could easily hear the sound of air sliding over the feathered wing surfaces. The gulls' wings made a dry sound, very different from the wet, spattering wing beat of a cormorant that had just risen from the water, and whose pre-cipitate flight down the cove sounded like a wet dog shaking himself.

Often, as I lay there, I could hear the high, peeping whistle of an osprey, and opening my eyes, would see him coming down along the inner shore of the island. I think a pair of them had a nest somewhere up north of the island; when they carried fish, they were always going north.

And then there were the sounds of other,

smaller birds — the rattling call of a kingfisher that perched, between forays after fish, on the posts of the dock; the call of the phoebe that nested under the eaves of the cabin; the redstarts that foraged in the birches on the hill behind the cabin and forever, it seemed to me, asked each other the way to Wiscasset, for I could easily twist their syllables into the query, "Which is Wiscasset? Which is Wiscasset?"

Sometimes the still water of the passage would be rippled, then broken, by the sleek, round head of a seal. Swimming up-current, his nostrils and forehead protruding, his passage sent diverging ripples running in silken V's toward the opposite shores. After looking gravely about him with soft, dark eyes, surveying for a moment the world of sun and air, the seal would disappear as silently as he had come, returning to the soft green lights, the seaweeds streaming from sunken rocks, the little silver gleams of fleeting fishes. There is always something of mystery about these mammals of the sea. Akin to ourselves in most of the biological processes, warm-blooded, possessing a hairy covering, suckling their young, yet they are at home in an element to which we can make only the briefest of visits.

Sometimes I would watch the island from

the hill that sloped up from the water line to a wooded crest from which could be seen the cove and all the outlying islands. It was fun to climb the hill, carpeted so thickly with gray-green reindeer moss, studded with pine and spruce and low-growing juniper. On the sunny slopes the moss was so dry that it crunched underfoot like very cold snow, but in the deep shade it was soft and spongy. Beardlike tufts of the strange Usnea moss or old-man's-beard hung from the pines, a suggestion that the beautiful parula warbler might be about, for the parulas nest in pendant clumps of this moss.

And indeed the woods there on the hillside were bright with the moving, flitting forms of many warblers — the exquisite powder-blue parula with his breast band of orange and magenta; the Blackburnian, like flickering flames in the spruces; the myrtle, flashing his yellow rump patch. But most numerous of all was the trim little black-throated green warbler, whose dreamy, nostalgic song drifted all day long through the woods, little wisps of song lingering like bits of fog in the tree tops. Perhaps because I so invariably heard it in those woods, when I now recall the song in memory, it always brings with it a vivid picture of that sunny hill splashed with the dark shadows of the evergreens, and the scent of all the heady, aromatic, bitter-sweet fragrances compounded of pine and spruce and bay-berry, warmed by the sun through the hours of a July day.

[1947]

Mattamuskeet:
A National Wildlife Refuge

In 1946 Rachel Carson submitted a plan to her Fish and Wildlife superiors for a twelve-part series highlighting the national wildlife refuge system. These Conservation in Action *booklets would serve not only as guides to individual refuges, but also as a forum for public education in ecology. The series gave Carson as editor an opportunity to design a model for Service publications.*

To do the necessary research, Carson planned visits to the selected refuges — Chincoteague, Parker River, Mattamuskeet, Bear River, Red Rocks Lake, and National Bison Refuge — beginning in 1946. This project afforded her the first and most extensive travel opportunity she would ever have, and it was the only time in her life she worked as a professional without the encumbrance of family.

Carson chose to feature in the series three waterfowl refuges on the Atlantic flyway. The southernmost was Mattamuskeet off Pamlico Sound in eastern North Carolina, notable for protecting the endangered whistling swan, which she visited with her friend and colleague artist Kay Howe in February 1947.

One morning Carson rose before dawn to walk out along the canal hoping to see the swans before they rose for the day's foraging. Geese flew off over her head "so close," she wrote in her notebook, "that I could hear the sound of their wings." Her sensory impressions of the waterfowl of Mattamuskeet, along with her acute observations of behavior and habitat, found their way into the text of this number four in the Conservation in Action *series, an ecological classic among government wildlife publications.*

[. . .] Mattamuskeet — the rhythmic softness of the Indian name recalls the days when tribes of the Algonquin roamed the flat plains of the coast and hunted game in deep forests of cypress and pine. The Indians are gone, leaving few traces upon the land they once knew. Much of the forest as the Indians knew it is gone, too, but even today some of the wildest country of the Atlantic coast is to be found in this easternmost part of the Carolina mainland — the area bounded by Albemarle Sound on the north and Pamlico Sound on the east and south. Here, in this coastal region, are dense woods of pine, cypress, and gum; here are wide, silent spaces where the wind blows over seas of marsh grass and the only living things are the birds and the small, unseen inhabitants of the marshes.

The Mattamuskeet National Wildlife Refuge includes about 50,000 acres of land and water in this Carolina coastal country, in the county of Hyde. The dominant geographic feature of the refuge is Lake Mattamuskeet — a shallow, sluggish body of water more than 15 miles long, 5 or 6 miles across, and some 30,000 acres in extent. Being little more than 3 feet deep anywhere, the lake is stirred deeply by the winds and its waters are usually muddy. Silt-filled waters support little plant life, and so the best feeding grounds for the waterfowl are not in the open lake but in its surrounding marshes. Cypress trees form most of the northern border of the lake, but its eastern and southern shores pass into low swamplands.

Try to learn the origins of this vast inland lake and at once you stumble upon a collection of local legends in which it is hard to separate fact from fiction. Of all the stories of the genesis of Mattamuskeet, local opinion divides its support between two. According to one story, the Indians long ago set fires in the peat bogs, fires that burned so long and deeply that a huge, saucer-like depression was formed. This caught the rains and the drainage water, creating a lake.

The other story has it that a shower of giant meteors once struck the Carolina coastal plain, the impact of the largest ones digging out the beds of Lake Mattamuskeet and the smaller, but otherwise similar, Lakes Alligator, Pungo, and Phelps that lie northwest of Mattamuskeet. [. . .]

Whistling swans are the most spectacular birds to be seen at Mattamuskeet. With their wing spread of 6 to 7 feet, they are the largest of all North American waterfowl except the related trumpeter swan, which is now reduced to less than 400 birds in the United States.

The whistling swans arrive at Mattamuskeet sometime in November, remain several months, and usually in February begin their northern migration. When they leave Mattamuskeet, they have a trip of 2,500 to 3,500 miles before them, for most of them breed north of the Arctic Circle. The species winters on the Atlantic coast, principally between Maryland and North Carolina, and also on the Pacific coast from southern Alaska to southern California.

A large flock of swans is noisy and their voices are a typical winter sound on the refuge. The mingled chorus of swan voices is something like the sound of geese, although somewhat softer. The name "whistling swan" is given because of a single high note sometimes uttered — a sound that suggests a woodwind instrument in its quality. The trumpeter has a deeper, more resonant voice because of an anatomical peculiarity — the windpipe has an extra loop. Trumpeters are never found on the Atlantic coast, however.

After a long history of persecution by man, all wild swans now enjoy complete protection in the United States, Alaska, and Canada. As though sensing this security, the swans at Mattamuskeet show very little fear of people and allow themselves to be approached much more closely than the geese. Five to ten thousand swans usually winter here, feeding in shallow water areas about the southern and eastern shores of the lake. It is possible to see a flock of 500 swans at one time, magnificent in their gleaming white plumage. Sometimes the swans feed or rest in family groups in which the young birds or cygnets may be identified by their grey color.

For the Canada geese of the Atlantic coast, Mattamuskeet is one of the chief wintering places, with a population of about 40 to 60 thousand of these handsome birds from November to the middle of March.

Magnificent though the swans are, the person who visits Mattamuskeet in midwinter is likely to come away with impressions of geese uppermost in his mind. Throughout much of the day, their wings pattern the sky above you. Underlying all the other sounds of the refuge is their wild music, rising at times to a great, tumultuous crescendo, and dying away again to a throbbing undercurrent.

Guided by the voices of the birds, you walk out along the banks of one of the canals about sunrise. A steady babble of goose voices tells you of a great concentration of the birds on the lake, probably off the end of the canal. At intervals the sound swells as though a sudden excitement had passed through the flock, and at each such increase in the sound a little party of birds takes off from the main flock and moves away to some favored feeding ground. As you stand quietly in the thickets along the canal, they pass so close overhead that you can hear their wings cutting the air, and see their plumage tinged with golden brown by the early morning sun.

The Mattamuskeet country is so famous for its geese that hunters come from great distances, and rent shooting blinds from farmers of the region or in the managed hunting areas operated on the refuge. In the 1946–47 season, the total kill of geese within these managed areas was 868. Large numbers are shot also in the surrounding countryside, but exact figures are not available.

A large majority — probably three-fourths — of the Mattamuskeet geese breed along the eastern shores of Hudson Bay, smaller numbers in the Maritime Provinces.

The ducks that winter at Mattamuskeet are largely the marsh or dabbling ducks — the shallow-water feeders. Pintails are the commonest of these, and it is a beautiful sight to see 10,000 or more of these graceful ducks wheeling above the marshes. Small flocks of wigeons appear in spring along the lake road. Black ducks, green-winged teal, mallards and blue-winged teal spend the winter here in varying numbers, from a few hundred to a few thousand.

Most of the ducks found in winter from Delaware Bay south nest in the prairie provinces of Canada or in the flat country of the

Dakotas and Minnesota. All of this country is subject to periodic droughts; then many ponds and marshes dry up, few ducks nest successfully, and few ducklings survive to join the fall flights south.

The bird clubs of North Carolina and surrounding States have made frequent visits to Mattamuskeet ever since the refuge was established. So many birds may be seen in the thickets or along the canals within a few hundred feet of the lodge that it is unnecessary for older members or others unable for strenuous exercise to go far afield. More than one person confined to a wheel chair, who had believed his days of field ornithology behind him, has been brought to Mattamuskeet for a satisfying and refreshing experience.

To gain the best vantage points for observing swans, geese, or ducks, it is worthwhile to hike out along the remnants of the former canals that here and there extend in long, densely overgrown peninsulas into the lake. Sometimes this will bring into view thousands of geese resting on the water. Concentrations of swans on feeding grounds along the south shore of the lake can sometimes be spotted from the highway, and can then be approached on foot within good binocular or camera range. All cultivated fields of the area should be watched for large flocks of geese.

The bird life of Mattamuskeet includes about 200 different species, with water birds and water-loving land birds predominating — less variety than is found in a more diversified country. Bird clubs visiting Mattamuskeet therefore may not compile a long list but see extremely large numbers of certain species, occasionally record a rarity, and have excellent opportunities for close observation of bird behavior.

Waterfowl are, of course, the chief winter attraction. Of these, swans, Canada geese, and the surface-feeding ducks find ideal conditions at Mattamuskeet. Diving ducks tend to go to the Swanquarter area. Marsh birds like herons are common: the great blue stays throughout the year, the American bittern is here in winter, the least bittern, green and little blue herons, and American egret are summer residents. Shorebirds, loons, and grebes find little suitable country for their habits and occur only in limited numbers.

The brown-headed nuthatch is a permanent resident, probably nesting on the islands of the lake or about the borders of the canals. In winter the wax myrtles are alive with myrtle warblers. Carolina wrens, chickadees, white-throated, fox, swamp, and song sparrows fill the winter thickets. Other winter residents or transients include the hermit thrush, ruby-crowned kinglet, pipit, horned lark, and cedar waxwing. The mockingbird is common throughout the year.

The most abundant of the summer warblers at Mattamuskeet is the prothonotary, with the prairie warbler also a common bird. Vireos, both white-eyed and red-eyed, are common in summer, as are wood thrushes and orchard orioles.

Observers of birds at Mattamuskeet over the years have marked up a number of unusual species, such as the white pelican, blue goose, white-fronted goose, Hutchins goose, black tern (a fall transient), European wigeon, black rail, and — as interesting stragglers from the west — the avocet and Arkansas kingbird.

What does the Mattamuskeet refuge do for the waterfowl that could not be done in the same area of wild country without management? This is a fair question, and its answer gives one of the chief reasons for establishing wildlife refuges in selected localities over the country.

The answer is this: by cultivating or managing the marshlands by scientifically tested principles, the land within the refuge is made many times as productive of natural foods as outside areas not under management.

Underlying and determining the character of the management activities are the great recurrent rhythms of nature. Moving over the marshlands as over a stage, the passing seasons bring the cyclic sweep of two great series of events, one in the animal world, the other in the world of plants. The two cycles are directly related. In the spring the marshes that have been brown and desolate come alive with fresh green shoots of plants like the sedges, bulrushes, and salt grass. Spring yields to summer, the hot sun is over the land, the plants grow, flower, mature their seeds. By the time autumn begins to paint the leaves of the gums and the swamp maples, the marshes are loaded with food — the roots, seeds, and shoots of the plants that waterfowl eat.

Now the fall migrations of the birds — the sweep of the other, the animal cycle — fill the marshlands with ducks, swans, and geese come down from the north. Here in the marshes they find the food they must have if they are to survive the winter.

By late winter or early spring the food supplies are exhausted. But once more the urge to migrate is stirring among the waterfowl, and soon the marshes are left empty. In the stillness and heat of summer the recuperative powers of nature set to work to build up new food supplies.

To get the largest possible production of waterfowl foods out of the marshes at Mattamuskeet, the manager operates the refuge with certain aims in mind. Among the most important, he must keep down the brush that is forever moving into the marshes. Geese, swans, and ducks feed in marshes but not in thickets, so every foot invaded by the fast-growing brush is a corresponding loss of waterfowl pasture. Today at Mattamuskeet you can see hundreds of acres of productive marsh which have been won back from the thickets by burning, disking, and cutting.

Control of the water level is another method used by the refuge manager to increase the production of food plants. In the spring he lowers the water by manipulating the gates in the canals that lead from the lake to Pamlico Sound, about 8 miles distant. This lays bare extensive areas where 3-edge, 4-square, and other food plants can grow. In the fall the gates are closed, and the marsh areas flooded to serve the food plants in the way the birds prefer — under a few inches of water.

By late January or early February, most of the natural marsh food has been eaten. The thousands of birds that remain must have food to fuel their bodies on the long spring migration. This is a season of busy activity on the refuge. Crews of men move out into the marshes, starting fires in the marsh grass. Keeping the fires carefully under control, many hundreds of acres are burned. Less than a week later, new green shoots are coming up all over the marsh. Within ten days the geese have moved in to harvest this new food supply.

By thus coordinating the management of the refuge with the natural cycles of plant and animal life, the Fish and Wildlife Service has developed Mattamuskeet to the point where it now supports much larger flocks of waterfowl than came to this region in former years.

Part Two

In the decade that began with the publication in 1941 of Under the Sea-Wind, *Carson's lyrical study of life in the open sea, and ended with the appearance of* The Sea Around Us *in 1951, Carson produced some of her most distinguished writing. The latter book, a monumental synthesis of the science of oceanography, catapulted her to international fame, provided a measure of fortune, and enabled her to leave the government and devote herself to her writing.*

Reticent at first about speaking in public, Carson eventually grew more self-assured in her role as public figure and used these occasions to promote natural history as a way of understanding the world. Her major themes — the timelessness of the earth, the constancy of its processes, and the mystery of life — are found over and over in the body of her writing, but they had a special freshness and intimacy when Carson spoke them aloud. Carson also used the opportunities offered by her many awards to speak out against the iconoclasm of science, and to urge the commonality of values shared by all those who endeavored to unfold the wonders of nature.

In 1952, Carson resigned from the U.S. Fish and Wildlife Service. No longer encumbered by government restrictions, she began to express openly her views on the politics of conservation and spoke out too for wilderness preservation.

Two of the selections in Part Two include references to the anxieties of life in the atomic age and to Carson's concern that with the atom bomb humankind had achieved the power to alter the natural world, even to destroy it. It was a truth which ultimately impelled her choice of subjects and lay at the heart of her anger at our despoiling arrogance.

⌁ 8 ⌁

[ca. 1942]

Memo to Mrs. Eales on
Under the Sea-Wind

Literary notice for Under the Sea-Wind, *Carson's first book on the life of the ocean, was prematurely cut short when the Japanese bombed Pearl Harbor barely a month after the book's publication. Eager nevertheless to assist Simon and Schuster's efforts to publicize her book, Carson completed an author questionnaire at the request of a Mrs. Eales in the publisher's marketing department. In this synopsis, Carson describes with exceptional candor how she came to write the book, how she conceived its parts and main characters, and what qualities made her approach to ocean life unique.*

Although Under the Sea-Wind *sold fewer than two thousand copies before it went out of print in 1946, the book was given new life when Oxford University Press republished it in*

1952 and it justly assumed a place on the New York Times *best-seller list along with* The Sea Around Us.

Background of the Book

It isn't at all surprising that I should have written a book about the sea, because as long as I can remember it has fascinated me. Even as a child — long before I had ever seen it — I used to imagine what it would look like, and what the surf sounded like. Since I grew up in an inland community, where we hadn't even a migrating seagull, I had to wait a long time to have my curiosity satisfied. As a matter of fact, it wasn't until I had graduated from college and gone to the famous Marine Biological Laboratory at Woods Hole, Massachusetts, that I saw the ocean. There, too, I began to get my first real understanding of the real sea world — that is, the world as it is known by shore birds and fishes and beach crabs and all the other creatures that live in the sea or along its edge. At Woods Hole we used to go out in a little dredging boat and steam up and down Vineyard Sound or Buzzards Bay. After a time, with much violent rocking of the little boat, the dredge would be pulled up and its load of sea animals, rocks, shells, and seaweeds spilled out on the deck.

Most of these animals I had never seen before; some I had never heard of. But there they were before me, dripping with sea water and perhaps clinging to a piece of rock or shell or weed that they had brought up from their home down there on the bottom of the sound. Probably that was when I first began to let my imagination go down through the water and piece together bits of scientific fact until I could see the whole life of those creatures as they lived them in that strange sea world.

In a way, *Under the Sea-Wind* had its beginning about six years ago, when I happened to write a short essay on life in the sea. A friend who read it suggested that I send it to the *Atlantic*. At first I didn't take the suggestion very seriously, for I let the essay lie in my desk for about a year; but finally I polished it up a little and sent it to the *Atlantic*. In due time a letter of acceptance came. A few weeks after the essay — which was called "Undersea" — had been published, I received a letter from Mr. Quincy Howe, who is editor-in-chief of the firm of Simon and Schuster. Mr. Howe said he had enjoyed the undersea article, and wondered if I was planning a book on the same general subject; if so, would I care to talk it over with him? As a matter of fact, I had never seriously considered writing a book, but naturally that letter put ideas in my head. I went to New York and we talked over a general plan for a book which would give the non-biologist a true picture of life in the sea. It was left that when I got around to writing such a book, the firm would like to consider it for publication. Actually, it was nearly two years later that the definite plan of the present book took form in my mind and I began to write, doing it all during my evenings and Saturday afternoons and Sundays. After I had written the first section the publishers signed a contract for publication of the book, and from that time on the writing went a lot faster, because a deadline had been set and I was writing under pressure, which sometimes isn't a bad thing.

General Plan and Viewpoint of Book

I believe that most popular books about the ocean are written from the viewpoint of a human observer — usually a deep-sea diver or sometimes a fisherman — and record his impressions and interpretations of what he saw. I was determined to avoid this human bias as much as possible. The ocean is too big and vast and its forces are too mighty to be much affected by human activity. So I decided that the author as a person or a human observer should never enter the story, but that it should be told as a simple narrative of the lives of certain animals of the sea. As far as possible, I wanted my readers to feel that they were, for a time, actually living the lives of sea creatures. To bring this about I had first, of course, to think myself into the role of an animal that lives in the sea. I had to forget a lot of human conceptions. For example, time measured by the clock means nothing to a shorebird. His measure of time is not an hour, but the rise and fall of the tides — exposing his food supply or covering it again. Again, light and dark may mean merely the difference between the time when you are relatively safe and the time when an enemy can find you easily. All these adjustments in my thinking had to be made; and in writing the book I was successively a sandpiper, a crab, a mackerel, an eel, and half a dozen other animals. Hardest of all, I had to get the feel of a world that was entirely water.

I very soon realized that the central character of the book was the ocean itself. The smell of the sea's edge, the feeling of vast movements of water, the sound of waves, crept into every page, and over all was the ocean as the force dominating all its creatures.

In order to give a fairly complete picture of

sea life, I divided the book into three parts, one to picture the life of the shore, one for the open sea, and one for the deep abyss. In each of these parts, or books, I told the life story of one particular animal.

Book I — Edge of the Sea

Almost everyone knows the sea beach in a general way. Unfortunately, though, most people stay within sight of the piers and boardwalks of a resort beach, and never become acquainted with the animals of the beach, except for the few whose remains may be found in the litter along the high-tide mark. I always seek out the wild sections of beach that are usually to be found a few miles above or below a resort. One particularly lovely stretch of wild ocean beach in North Carolina forms the setting for most of the chapters about the shore. It is a beach that is separated from the mainland towns by a wide sound, fringing one of those narrow strips of outer land which the Carolinians call "banks." I have visited that beach in spring and fall to watch the comings and goings of the shorebirds. I have spent hours on end among the dunes or on the beach, saturating myself with the sounds of water and the feel of hot sun and blowing sand. I have watched the fiddler crabs and ghost crabs, and, in autumn, seen the mullet fishermen draw their seines on the beach. This was the background for the story of a bird that almost everyone who has visited the beach has seen as it runs along at the edge of the waves — a special kind of sandpiper called a sanderling. I chose the sanderling as the main character of the shore section because of its fascinating life story. The sanderling is one of the bird tribe's long-range migrants. I believe that few of the people who like to watch the sanderling on the beach have any idea of the hardships these birds endure, or the long and hard flights they make. Some of them actually travel more than eight thousand miles every spring and return the same distance every fall. These little birds winter as far south as Patagonia, at the extreme southern end of South America, and in the spring they migrate northward, most of them beyond the Arctic Circle, and some to within a few miles of the North Pole. This seems a strange place to choose to rear their young, but many of the shore and ocean birds nest in the Arctic. Probably they are obeying some instinct inherited from forgotten generations of ancestors. We see the sanderlings in the spring as they are migrating up along our coastline, then, about May and June in Maryland and

Virginia, all but a few immature birds disappear. This is during the period when the adult birds are nesting on the Arctic tundras. When they first arrive in the Arctic the snow and ice have not melted, food is scarce, and late-season blizzards may take a heavy toll of life. Eventually, spring comes even to these frozen tundras, the birds prepare their nests and lay their eggs, and the young are hatched. There are many enemies abroad on the tundra. Some of these are the large snowy owls, the foxes, and hawk-like birds of the gull tribe known as jaegers. After the chicks have hatched, the mother sanderling takes great precautions to hide the egg shells, so that enemies will not be led to the nest by them. Usually she leads the young away from the nest when they are only a few days or even hours old, if she has been frightened by marauders. Very quickly, however, the young become able to take care of themselves. As soon as they are no longer needed, the old birds leave for the south. The young remain behind until their wing feathers have grown strong enough for the long journey down across the two Americas. By late July the older sanderlings are seen on our beaches again, and a few weeks later we begin to see young birds.

This is the story that I have told in the first

section, against a background of Carolina beach and Arctic tundra.

Book II — The Gull's Way

The central character of the second section is another long-distance migrant, but this time a fish. In this section, which pictures the strange world of the open sea, I have written the biography of a mackerel, beginning, as biographies usually do, with the birth of my central character. There could scarcely be any stranger place in which to begin life than the surface waters of the open sea. Yet these waters are a sort of nursery where literally hundreds of kinds of sea creatures deposit their eggs, and where the young get their start in life. Parenthood in the sea is a relatively simple matter, for as a usual thing the parents do not care for their young and probably never even see them.

The open sea *is* a strange place for anything so fragile as a mackerel egg to be set adrift: just sky and water, and great silences, but teeming, incredibly abundant life. In the first place there are the eggs of all sorts of animals — fishes, crabs, shrimps, clams, worms, starfish, and the like. From all these eggs larvae or young animals are hatching. Almost immediately, each larva is on its own resources. It

101

begins to swim about and seek food, eating almost anything that is small enough to take into its mouth, or to overpower and swallow. All sorts of enemies of young fishes prowl through these surface waters: small jellyfish with enormous appetites, little, transparent worms with sharp, biting jaws, schools of small fishes that eat smaller fishes, and larger fishes that eat them. Just to give an idea of some of the hazards of sea life: a full-grown mackerel may produce half a million eggs in a season, or a large cod may shed three or four million. But the destruction of the young is so enormous that, on the average, only two young mackerel or cod will survive out of all the potential offspring produced by the mother fish during her whole life. This ceaseless ebb and flow of life — the constant destruction of individuals contrasted with the survival of whole species — is one of the most impressive spectacles which the sea presents.

As the young mackerel grows rapidly during the first months of life, sea animals that were once deadly enemies become his prey as he, too, joins the ranks of sea hunters. After spending the summer in a sheltered New England harbor, he and other young mackerel wander out into the open sea again. There new and larger enemies await them: fish-eating birds, swordfish, tunas, and fishermen. In the concluding chapter of this series, I described the setting of a mackerel seine from the viewpoint of a fish — something that I do not believe has been done before.

In many ways, I found this section the hardest to write, and so I get a good deal of satisfaction out of the fact that most reviewers and readers seem to like it best. I believe it was hard because of the endless waste of waters — no fixed points around which to orient one's characters. I said a few minutes ago that I really lived the things I wrote about, and I don't mind admitting that I was very thankful to climb out on dry land in beginning the concluding section.

Book III — River and Sea

For the last section of the book, I had left the gently sloping sea bottom from the tidelines out to the edge of the continental shelf, and the deep Atlantic abyss. There was one fish whose migrations include all that varied undersea terrain — the eel. I know many people shudder at the sight of an eel. To me (and I believe to anyone who knows its story) to see an eel is something like meeting a person who has travelled to the most remote and wonderful places of the earth; in a flash I see a vivid picture of the strange places that eel has been — places which I, being merely human, can never visit.

Every eel that lives along our Atlantic coast began life in the distant Sargasso Sea. It lived, at first, so far below the surface that only the faintest blue haze ever penetrates there. For the most part, the water in which the baby eels are born is eternally dark and still and cold. The pressure is so great that it would instantly crush our unaccustomed bodies to nothing. All about the baby eels are the strange animals that live permanently in the abyss. Many of them carry their own lights, perhaps to help them see their way about in the darkness and find food.

As the young eels grow they work up toward the surface, and as they move up the light becomes stronger. By this time they look like tiny willow leaves, flat, oval, and transparent. In a few months' time, they begin their thousand-mile journey toward the American coast. At first, probably, they are carried along by the ocean currents; later, they must swim independently. But here is the really remarkable part of the story. In the Sargasso, the young of eels from America mingle with the young of European eels, for the eels from all the European Atlantic coast make the long westward crossing to spawn in the Sargasso. But although many of the two species of young are intermingled during their first weeks or months of life, soon after the migration begins the travelers separate. They form two great bands, one company proceeding westward toward America, the other eastward to Europe. The two kinds of eels are so similar that a scientist can distinguish them only by counting the number of vertebrae in the backbone, but the little eels themselves never make a mistake. They always return to the continent from which their parents came.

In the spring the young eels begin to arrive in our coastal waters. They are, by this time, a little more than a year old, but they are no longer than a man's finger and so transparent that one could read print through their bodies. They move into the bays and river estuaries, and some begin to ascend the rivers and streams. It is thought that the young males remain in salt or brackish water, and that it is only the females that ascend the fresh-water streams. There they live for 8, 10, or 12 years before they reach physical maturity. Then the awakening of some race instinct causes them to begin a downstream migration. This happens in the fall of the year. Usually the eels migrate at night, and apparently dark, stormy nights are times of large movements of eels. In the estuaries of the rivers the migrating females are joined by the males and together they enter the sea and pass out through the coastal waters. Fishing boats take a few; then the eels completely disappear from sight and never are seen again. We know, though, that they returned to their birthplace a thousand miles out in the Atlantic, because, very early in the spring, the eggs of the new generation of young can be found there. Evidently the old eels die after they spawn, for they never return to the coast. They begin and end their lives in the deep abyss.

General

Each of these stories seems to me not only to challenge the imagination, but also to give us a little better perspective on human problems. They are stories of things that have been going on for countless thousands of years. They are as ageless as sun and rain, or as the sea itself. The relentless struggle for survival in the sea epitomizes the struggle of all earthly life, human and non-human. As one reviewer said: "Our own battles for existence seem less a matter for dismay and more a simple reason for fortitude when compared in the mind with the ceaseless ebb and flow of life and death that are under all the sea winds."

∿ 9 ∿

[1949]

Lost Worlds: The Challenge of the Islands

Desperately in need of money to support her family — her mother, and her two young nieces — and to further her publishing ambitions for the book that would become The Sea Around Us, *Carson engaged Marie Rodell, a New York literary agent, in the spring of 1948. One of their publishing strategies was to sell individual chapters as soon as Carson completed them. The chapter dealing with the creation of oceanic islands was one of the most scientifically challenging and one Carson judged from the outset could stand alone as an independent essay.*

The story of how islands are formed and inhabited went through many versions before Carson was satisfied. Her research on island evolution was aided by F. Raymond Fosberg, a tropical botanist at George Washington Univer-

sity and the Smithsonian's National Museum, and a world expert on atoll formation. Fosberg read an early draft of Carson's chapter which he later described as "the finest account of the creation and colonization of an oceanic island" he had ever read.

This version, titled "Lost Worlds," was published in the magazine of the D.C. Audubon Society, The Wood Thrush, *edited by Carson's friend Shirley Briggs, in the spring of 1949. It is distinguished by Carson's undisguised anger at the human destruction of the rare ecology of island habitats, her advocacy of island ecosystem preservation, and her delight in the mysterious processes of species migration to distant Atlantic atolls. Although a later version published in the* Yale Review *was awarded the Westinghouse Science Writing Prize from the American Association for the Advancement of Science, "Lost Worlds" brought Carson favorable notice from the small but influential Washington, D.C., science community whose support was crucial to her expanding literary career.*

Dr. Ernst Mayr of the American Museum of Natural History recently compiled a list of all species of birds known to have become extinct anywhere in the world during the past two centuries. This was his score: on all continents combined, eight species; on islands, at least ninety-two — probably more than a hundred when all reports from war areas are in.

This report epitomizes the tragedy of island life that is playing out what may well be its last act before our eyes. Each of the ninety-two species named by Dr. Mayr represents a loss that will never be replaced. For most of these island species have been created once, and only once, in all the world, the products of the slow processes of the ages. Destroyed by man's careless abuse of the most delicately balanced environment in the world, an oceanic island, they are lost forever.

The problem of the islands is not one that can be put off until later; it is not one that will solve itself if we adopt a comfortable policy of laissez faire. Our own generation is in all probability the last that will have an opportunity to save any of the original island faunas and floras. The Atlantic islands, whose discovery and colonization began back in the sixteenth century, were despoiled so long ago that we scarcely realize what was lost. The islands of the Indian Ocean and parts of the Pacific came in for their turn a little later. The immense distances of the vast Pacific, the remoteness of many of its islands from the routes of whalers and traders, for a time saved some of the Pacific islands, but not for long. Today there are in all the world only a few islands whose original life remains.

Islands present a conservation problem that is absolutely unique, a fact that is not generally realized. This uniqueness stems from the nature of the island species, and from the delicately balanced relationships between island animals and plants and their environment. And going back still farther, these things are related to the origin of the islands themselves, and to the amazing manner in which they acquired their faunas and floras.

The islands of the deep ocean, far from the continents, are the products of an extraordinary process of earth-building. With few exceptions, they are the result of the violent, explosive, earth-shaking eruptions of submarine volcanoes, working perhaps for thousands or millions of years. In eruption after eruption the mass of an undersea mountain takes form on the floor of the ocean, builds up toward the surface, emerges as an island.

On their first emergence from the sea, these islands are bare, harsh, and repelling beyond human experience. No living creature moves over their volcanic hills; no plants cover their naked lava fields. By what miracle are these islands, isolated by hundreds or thousands of miles from other land, transformed into forested hills and fertile valleys, bright with birds and stirring with life?

The stocking of the islands has been accomplished by the strangest migration in earth's history — a migration that began long before man appeared on earth and must still be continuing, a migration that seems more like a series of cosmic accidents than an orderly process of nature. Little by little, riding on the winds, drifting on the currents, or rafting in on logs, floating brush, or trees, the plants and animals that are to colonize them arrive from the distant continents.

So deliberate, so unhurried, so inexorable are the ways of Nature that the stocking of an island may require thousands or millions of years. It may be that no more than half a dozen times in all these eons does a particular form, such as a tortoise, make a successful landing upon its shores. To wonder impatiently why man is not a constant witness of such arrivals is to fail to understand the majestic pace of the process.

Yet we have occasional glimpses of the method. Natural rafts of uprooted trees and matted vegetation have frequently been seen adrift at sea, hundreds of miles off the mouths of such great tropical rivers as the Congo, the Ganges, the Amazon, and the Orinoco. Such rafts could easily carry an assortment of insect, reptile, or mollusk passengers. Some of these involuntary passengers might be able to withstand long weeks at sea; others would die during the first stages of the journey. Probably the best adapted for travel by raft are the wood-boring insects, which, of all the insect tribe, are most commonly found on oceanic islands. The poorest raft travelers must be the mammals, yet even a mammal might cover short inter-island distances.

No less than the water, the winds and the air currents play their part in bringing inhabitants to the islands. With special nets and traps, scientists have now collected from the upper atmosphere many of the forms which inhabit oceanic islands. Spiders, whose almost invariable presence on these islands is an intriguing problem, have been captured nearly three miles above the earth's surface. Airmen have passed through great numbers of the white, silken filaments of spiders' "parachutes" at heights of two to three miles. At altitudes of 6,000 to 16,000 feet, and with wind velocities reaching 45 miles an hour, many living insects have been taken. At such heights and on such strong winds, they might well have been carried hundreds of miles. Seeds have been collected at altitudes up to 5,000 feet. Among those commonly taken are members of the Composite family, typical of oceanic islands.

The wide-ranging birds that visit islands of the ocean in migration may also have a good deal to do with the distribution of plants, and perhaps even of some insects and minute land shells. From a ball of mud taken from a bird, Charles Darwin raised eighty-two separate plants, belonging to five distinct species! Many plant seeds have hooks or prickles, ideal for attachment to feathers. Such birds as the Pacific Golden Plover, which annually flies from the mainland of Alaska to the Hawaiian Islands and even beyond, probably figure in many riddles of plant distribution.

Isolated from the great mass of life on the continents, with no opportunity for the cross-breeding which tends to preserve the average, to eliminate the new and unusual, island life has developed in a remarkable manner. On these remote bits of earth, Nature has excelled in the creation of strange and wonderful forms. As though to prove her incredible versatility, almost every island has developed species which are endemic; that is, they are peculiar to it alone, and are duplicated nowhere else on earth.

The strange plants and animals of the Galapagos Islands — giant tortoises, black, amazing lizards that hunted their food in the surf, birds in extraordinary variety — moved Charles Darwin, years after his visit to the islands, to write in reminiscence: "Both in space and time we seem to be brought somewhat near to that great fact — that mystery of mysteries — the first appearance of new beings on earth."

Of the "new beings" evolved on islands, some of the most striking examples have been birds. In some remote age before there were men, a small, pigeon-like bird found its way to the island of Mauritius, in the Indian Ocean. By processes of change at which we can only guess, this bird lost its power of flight, developed short, stout legs, grew larger until it reached the size of a modern turkey. Such was the origin of the fabulous Dodo, which did not long survive the advent of man on Mauritius. New Zealand was the sole home of the Moa, an ostrich-like bird that stood twelve feet high. Moas had roamed New Zealand from the time of the Pliocene, but they died out soon after the arrival of the Maoris.

Besides the Dodo and the Moa, other island forms have tended to become large. The loss of wing use and even of the wings themselves (the Moa had none) is a common result of insular life. Insects on small, wind-swept islands lose the power of flight.

The Galapagos Islands have a flightless cormorant. There have been at least fourteen species of flightless rails in the islands of the Pacific alone.

One of the most interesting and engaging characteristics of island species is their extraordinary tameness, a lack of sophistication in dealings with the human race which even the bitter teachings of experience do not quickly alter. When Robert Cushman Murphy visited the island of South Trinidad in 1913 with a party from the brig DAISY, terns alighted on the heads of the men in the whaleboat and peered inquiringly into their faces. Albatrosses on Laysan, whose habits include wonderful ceremonial dances, allowed naturalists to walk among their colonies, and responded with a grave bow to similar polite greetings from the visitors. When the British ornithologist David Lack visited the Galapagos Islands, a century after Darwin, he found that the hawks allowed themselves to be touched, and the flycatchers tried to remove hair from the heads of the men for nesting material. "It is a curious pleasure," he wrote, "to have the birds of the wilderness settling upon one's shoulders, and the pleasure could be much less rare were man less destructive."

But man, unhappily, has written one of his

blackest records as a destroyer on the oceanic islands. He has seldom set foot on an island that he has not brought about disastrous changes. He has destroyed environments by cutting, clearing, and burning; he has brought with him as a chance associate the nefarious rat; and almost invariably he has turned loose upon the islands a whole Noah's Ark of goats, hogs, cattle, dogs, cats, and other non-native animals and plants. Upon species after species of island life, the night of extinction has fallen.

In the world of living things, it is doubtful whether there is a more delicately balanced relationship than that of island life to its environment. In the midst of a great ocean, ruled by currents and winds that rarely shift their course, climate changes little. There are few natural enemies, perhaps none. The harsh struggle for existence that is the normal lot of continental life is softened on the islands. When this gentle pattern of life is abruptly changed, the island creatures have little ability to make the adjustments necessary to survival.

Ernst Mayr tells of a steamer wrecked off Lord Howe Island east of Australia in 1918. Its rats swam ashore. In two years they had so nearly exterminated the native birds that an islander wrote: "This paradise of birds has become a wilderness, and the quietness of death reigns where all was melody."

On Tristan da Cunha, all of the unique land birds that had been evolved there in the course of the ages were exterminated by the hogs and the rats. The native fauna of Tahiti and thousands of other Pacific islands is losing ground against the horde of alien species that man has introduced.

Most of man's habitual tampering with Nature's balance by introducing exotic species has been done in ignorance of the fatal chain of events that would follow. But in modern times, at least, we might profit by history. About the year 1513, the Portuguese introduced goats onto the recently discovered island of St. Helena, which had developed a magnificent forest of gumwood, ebony, and brazilwood. By 1560 or thereabouts, the goats had so multiplied that they wandered over the island by the thousand, in flocks a mile long. They trampled the young trees and ate the seedlings. By this time the colonists had begun to cut and burn the forests, so that it is hard to say whether men or goats were the more responsible for their destruction. But of the result there was no doubt. Even as early as the year 1880 the naturalist Alfred Wallace had to describe this once beautiful, forest-clad volcanic island as a "rocky desert," in which the fugitive remains of the original flora persisted only in the most inaccessible peaks and craters.

When the astronomer Halley visited the islands of the Atlantic about 1700, he put a few goats ashore on South Trinidad. This time without the further aid of man, the work of destruction proceeded so rapidly as to be nearly completed within the century. Today Trinidad's slopes are the place of a ghost forest, strewn with the fallen and decaying trunks of long-dead trees; its soft volcanic soils, no longer held by the interlacing roots, are sliding away into the sea.

The Hawaiian Islands, which have lost their native plants and animals faster than almost any other area in the world, are a classic example of the results of interfering with natural balances. Certain relationships of animal to plant, and of plant to soil, had grown up through the centuries. When man came in and rudely disturbed this balance, he set off a whole series of chain reactions.

Vancouver brought cattle and goats to the Hawaiian Islands, and the resulting damage to forests and other vegetation was enormous. Many plant introductions were as bad. A plant known as the pamakani was brought in many years ago, according to report, by a Captain Makee for his beautiful gardens on the island of Maui. The pamakani, which has light, wind-borne seeds, quickly escaped from the Captain's gardens, ruined the pasture lands on Maui, and proceeded to hop from island to island. The CCC boys once were put to work to clear it out of the Honouliuli Forest Reserve, but as fast as they destroyed it, the seeds of new plants arrived on the wind. Lantana was another plant brought in as an ornamental species. Now it covers thousands of acres with a thorny, scrambling growth — despite large sums of money spent to import parasitic insects to control it.

There was once a society in Hawaii for the special purpose of introducing exotic birds. Today when you go to the islands, you see, instead of the exquisite native birds that greeted Captain Cook, mynahs from India, cardinals from the United States or Brazil, skylarks from Europe, and titmice from Japan. Most of the original bird life has been wiped out, and to find its fugitive remnants you would have to search assiduously in the most remote hills.

One of the most interesting of the Pacific islands was Laysan, one of the far outriders of the Hawaiian chain, a tiny scrap of volcanic soil. It once supported a forest of sandalwood and fanleaf palms, and had five land birds, all peculiar to Laysan alone. One of them was the Laysan Rail, a charming, gnome-like creature no more than six inches high, with wings that seemed too small (and were never used as wings) and feet that seemed too large, and a voice like tinkling bells. About 1887, the captain of a visiting ship moved some of the Rails to Midway, establishing a second colony. This seemed a fortunate move, for soon thereafter rabbits were introduced on Laysan. Within a quarter of a century the rabbits had killed off the vegetation of the tiny island, reduced it to a sandy desert, and all but exterminated themselves. As for the Rails, the devastation of their island was fatal, and the last Rail on Laysan died about 1924.

Perhaps the Laysan colony could later have been restored from the Midway group had not tragedy struck there also. During the war in the Pacific, rats went ashore from ships and landing craft on island after island. They invaded Midway in 1943. The adult Rails were slaughtered. The eggs were eaten, and the young birds killed. The world's last Laysan Rail was seen in 1944.

The disruptive forces that had been operating for centuries throughout the Pacific were greatly accelerated by war. Some of the destruction was the direct result of bombing and artillery fire, but much of it was indirect. Ulithi Atoll in the Carolines was the home of a small rail, found nowhere else. The rail survived the early period of invasion, but perished when the taro swamps in which it lived were filled to make way for quonset huts. Large birds like albatrosses, shearwaters, and petrels often fell into abandoned foxholes and other steep-sided pits from which they could not escape, and starved. Planes killed thousands of birds, especially kinds that are active at night, like the Sooty Terns.

Out of the Pacific war, however, grew the first recognition of the conservation problem of the islands, and the first small beginnings of a constructive movement to salvage what remains. In 1946, the Pacific War Memorial was established. One of its purposes is the commemoration of lives lost in the Pacific by preserving, as living memorials, examples of original island life. Late in 1948 the Pacific War Memorial established a laboratory on the island of Koror, in the Palau Archipelago, to begin a study of conservation problems. The conservation crisis in the Pacific islands has also been the subject of several conferences sponsored by the National Research Council's Pacific Science Board. The advice of leading specialists on Pacific conservation problems has been sought by the Navy in connection with the Trust Territory of Micronesia.

There is still a chance to preserve some of the unique island life of the Pacific by establishing sanctuary areas comparable to our own National Parks and Wildlife Refuges. The first actual start on such a program has just been made. Late in 1948, the Navy turned over to the Pacific War Memorial two areas on Saipan as conservation reserves to commemorate the men who died in the fighting on this island. Between them, the two areas — Lake Susupe and Mt. Tapotchau — contain almost the only remaining vestiges of the original wildlife and forests of the island.

Lake Susupe, with its surrounding swamp, is the last stronghold in the world of one of Micronesia's most interesting birds, the Marianas Mallard. This bird has always been rare, and museums anywhere in the world that had a specimen counted themselves lucky. It was first described by scientists less than a century ago, from a single specimen in the Paris museum. It was found only on Guam, Tinian, and Saipan, and even there flocks of more than fifty or sixty birds were an unusual sight. It now seems to have disappeared from both Tinian and Guam, and probably not more than a score remain on Saipan. Under protection on Lake Susupe, conservationists hope that this remnant may build up enough to save the species from extinction.★

Mt. Tapotchau is in the high interior. Its jungled ravines and high ridges shelter most of what remains of the original forest of Saipan. Japanese farmers, clearing the island for sugar cane plantations, cut down much of the old forest elsewhere on the island, and war bombardments leveled the rest. Now the surviving native species on Mt. Tapotchau are threatened by the enemies of all island forests: burning, cutting, displacement by introduced plants, attack by insects and disease.

What can conservation areas accomplish? The species that are gone cannot be restored by any amount or kind of conservation work. As for those that remain, the example of the island of Lanai in the Hawaiian chain gives reason to hope that even badly damaged Pacific forests, and their associated life, could be brought back.

By about 1910, most of the forests and other vegetation of Lanai had been devoured by the cattle, sheep, goats, pigs, and deer which had been brought to the island over the years and allowed to run wild. Erosion on the

★ Ed.: The Marianas Mallard has been extinct in the wild since the 1970s. The last one died in captivity in 1981.

north end had become so serious that the island was literally blowing away. About that time George C. Munro was sent to the island to manage the Lanai Ranch. Munro, a conservationist by instinct, had the practical common sense to realize that a ranch could not be a paying proposition unless the cattle had something to graze on. He took drastic action. He had the wild cattle driven into corrals to augment the depleted ranch herds. Then he and his men declared a relentless shooting war on the wild pigs, goats, sheep, and deer. They built miles of fences and kept even the ranch cattle away from the mountain forests.

A quarter of a century later the botanist F. R. Fosberg went to Lanai to collect, as he expected, the few dying remnants of a once magnificent flora. Instead, he found that a miracle of restoration had taken place. Once more, the ridges and valleys of Lanai were covered with extensive forests of native trees. The erosion on the north end of the island had been halted. Preserved on Lanai, as in a museum, were several Hawaiian endemics now to be seen nowhere else in the world, among them an exquisitely fragrant gardenia, and a small mint restricted to a tract less than an acre, now its only habitat on earth.

Whether the efforts of the Pacific War

Memorial, the Pacific Science Board, and other conservation groups now at work in the Pacific have come in time, and have sufficient momentum, to achieve their aims, only the future will tell. As always in conservation problems, public ignorance and public apathy are the greatest obstacles to success. The degree of understanding and the amount of material support which are given these programs may well give the final answer to the challenge of the islands within our generation.

✌ 10 ✌

[1951]

New York Herald-Tribune
Book and Author Luncheon Speech

Rachel Carson was somewhat uncomfortable with the public role she assumed when The Sea Around Us *made her a literary celebrity. Unaccustomed to public speaking, she reluctantly agreed to appear at the* New York Herald-Tribune Book and Author Luncheon *after Irita Van Dorn, the irrepressible book review editor, invited her less than a month after* The Sea Around Us *had been published.*

Carson prepared a brief speech on the mystery and fascination of the sea, and to fill time, armed herself with hydrophone recordings of sounds made by shrimp, whales, and other fish in the sea's middle region that she had borrowed from the Woods Hole Oceanographic Institution. Her talk on the ancient evolution of the world's ocean and its life was a great success and ironically left her much in demand as a speaker, although she spoke so softly some had to strain to hear her.

Carson once remarked that she was always more interested in what she was going to write about next than in what she had written. Her remarks here indicate that she was already involved in research for her next book on the transition of life from sea to land.

People often seem to be surprised that a woman should have written a book about the sea. This is especially true, I find, of men. Perhaps they have been accustomed to thinking of the more exciting fields of scientific knowledge as exclusively masculine domains. In fact, one of my correspondents not long ago addressed me as "Dear Sir" — explaining that although he knew perfectly well that I was a woman, he simply could not bring himself to acknowledge the fact.

Then even if they accept my sex, some people are further surprised to find that I am not a tall, oversize, Amazon-type female. I can offer no defense for not being what people expect, but perhaps I might say a few words about why a woman, and only an average-size one at that, should have become a biographer of the sea.

I seem to have been born with a fascination

for the ocean. For years before I had ever seen it, I thought about it and dreamed about it and tried to picture what it would be like. I loved Swinburne and Masefield and all the other great poets of the sea. The stories I wrote for my classes in English composition often had a marine background. After I became interested in biology as a college student, I very naturally came to specialize in marine biology. At the Marine Biological Laboratory at Woods Hole I had my first prolonged contact with the sea. There I never tired of watching the tidal currents pouring through the Hole, and the waves breaking on Nobska Point after a storm. It was there, too, that I first discovered the rich scientific literature of the sea. But it is fair to say that my first impressions of the sea were sensory and emotional, and that the intellectual response came later.

Recently I have discovered that a great many other people feel just as I do about the ocean. A really overwhelming number of them have written to tell me so. During this past summer I have been traveling along the New England coast, gathering material for a new book. And I have been looking at people as well as at sea animals. I have been deeply impressed by what I saw. Everywhere there were people who simply sat — or stood —

and gazed out over the sea, without saying a word. Whatever they consciously thought, the spell that was cast over them was clearly written on their faces. I have been trying to analyze some of the reasons for this fascination.

The sea is a place where one gets a sense of the great antiquity of the earth. It seems changeless; but it is always changing. It links the dim beginnings of time with the present. The same sort of waves that we watch today must have rolled in from Paleozoic seas. I suppose that the surface waters of the ocean look much as they did half a billion or a billion years ago, when the first primitive forms of life stirred in it. And even some of our shores must look about as they did in that period, some 300 million years ago, when the first animals were coming out on land to take up a strange new life.

I was reminded of that last summer when I stood on one especially beautiful point on the rocky coast of Maine. We had come down to the point through an evergreen forest that had its own sort of enchantment. All its trees, the living and the dead, were hung with the silvery grey of mosses and lichens. But it was a foggy morning, and when we reached the rocks above the surf the mists lay between us and the forest, and all we could see were those

129

massive, primeval rocks and the sea. Except for our own presence, the scene might have been one of the closing periods of the Paleozoic Era. Some of the animals clinging to the walls of the tide pools might almost have been those early pioneers from the sea, that first came out on land back in Silurian time.

Now here is the particular magic of the sea. Exciting things are happening there today, just as they did millions of years ago. Evolution and the adaptation of creatures to new surroundings did not stop back in prehistoric time; they are still going on. That very day, only a few weeks ago, we saw hundreds of small, inconspicuous sea animals in the midst of a great experiment — the transition from a sea life to a land life.

These animals were small snails known as periwinkles. I am sure all of you have seen them on rocky coasts, between the tide lines. In some places you can hardly step without treading on the dingy grey shells of the common periwinkle. The periwinkles are now in the process of leaving the sea and turning into land snails. One by one, they are cutting the ties that bind them to the sea. Some of them have made more progress in this direction than others.

Here on our northern Atlantic coast there are three species of periwinkles. One of them is still almost completely marine. It lives down among the rockweeds where it is always wet or at least very damp. It lays its eggs on the weeds, and the young hatch out there and develop. Another species, called the common periwinkle, comes far up on the shore — as far as the waters of the high tide. It can stand a good deal of exposure to the air. In fact, it has developed a very simple sort of lung for breathing out of water. But it is still dependent on the sea, for it sheds its eggs into the water, and all the baby common periwinkles must spend the first period of their existence swimming about in the waters of the ocean.

The third species, called the rough periwinkle, is almost a land animal. Some of them live in crevices in the rocks where they are wet only by the spray of breaking storm waves. They can live without any contact with sea water for a week or more. Even in their method of reproduction they have cut their ties with the sea. The young of this species undergoes complete development within the body of the mother. They emerge as little snails exactly like their parents, ready for adult life. And so the three species of periwinkles give us a beautiful demonstration of the pattern of evolution, as it has been working out in the sea over the ages.

That is part of the fascination of the ocean. But most of all, the sea is a place of mystery. One by one, the mysteries of yesterday have been solved. But the solution seems always to bring with it another, perhaps a deeper mystery. I doubt that the last, final mysteries of the sea will ever be resolved. In fact, I cherish a very unscientific hope that they will not be.

A century is a very short time. Yet only a century ago men thought nothing could live in the deep waters of the ocean. They believed that, at most, there could be only a "few sparks" of life in the black waters of the oceanic abyss. Now, of course, we know better. In the year 1860 a surveying vessel was looking for the best route for the trans-Atlantic cable. When the sounding line was brought up from a depth of about a mile and a half, there were starfish clinging to the line. The same year a cable was brought up for repairs from the bottom of the Mediterranean. It was heavily encrusted with corals and other animals that evidently had been living on it for months or years. Such discoveries gave our grandfathers and our great-grandfathers their first proof that the floor of the deep sea is inhabited by living creatures.

Now, in our own time, another mystery of the sea is engaging the attention of scientists. This is the nature of the life of those strange, middle regions — far below the surface, but also far above the bottom.★

We had always assumed that these mid-depths were a barren, almost lifeless, Sahara of the sea. They lie beyond reach of even the strongest rays of the sun. And where there is no sunlight, no plants can live. So we assumed that food would be too scarce to support a very abundant animal population there.

Then about ten years ago came the discovery of immense concentrations of some living creatures, spread like a cloud over much of the ocean at a depth of a quarter of a mile or more. No one is sure just what these creatures are. As yet they have been "seen" only with the impersonal eye of echo sounding

★ Ed.: About the same time as Carson made this speech, scientists at Woods Hole were identifying and refining their understanding of "the scattering layers," or what she refers to here as "the middle region of the ocean," of which there are many. Although the exact population of fish within these layers depends on water temperature and the upwelling of deep ocean currents, scientists have found a wide variety of fish but not necessarily a large number of individual fishes. In spite of the fact that Carson's recordings were taken from these depths, they turned out not to be particularly noisy regions of the ocean or, unfortunately, especially abundant.

instruments. These instruments automatically record the depth of water under a moving vessel. They trace the contour of the ocean floor as a continuous line on a strip of paper. They also record, as traces or smudges on the paper, any solid objects, like schools of fish, that lie between the surface and the bottom. Hundreds of vessels have now found this layer of living creatures over the deeper parts of all oceans of the world. It has sometimes been called the "phantom bottom" of the sea, because people at first mistook it for shoals or sunken islands, and reported submerged land where none existed. Everyone agrees now that the layer is composed of living creatures. At night — in darkness — it moves up to the surface of the sea. But just before daybreak it descends again into deep water where light cannot follow it. Many small shrimplike creatures are known to do this. Also, some of the weird fishes of the deep sea come to the surface at night, like those Mr. Heyerdahl described so vividly in *Kon-Tiki.*

Scientists have tried to sample the layer with nets. You can never be sure, though, that your nets are catching everything. Perhaps the very creatures that are the key to the mystery are too swift to be caught. So the results are not very satisfying. Some people think the mystery creatures are shrimps — billions and billions of them. Others think they are fish or squid. If they should turn out to be something edible, the layer would represent an enormous food supply because of its almost ocean-wide dimensions. The answer to this enigma may come very soon, for a great many people are working on it.

One very common misconception about the sea was corrected by studies made during the Second World War. We always used to think of the deep sea as a place of silence. The idea that there could be sound under water had not entered most people's minds. Nor had the idea that fish or shrimp or whales had voices. When Navy technicians began listening for submarines during the war, they heard a most extraordinary uproar. In fact, the tumult of undersea voices was so great that whole fleets of submarines could have passed by undetected. Later, of course, means were developed for filtering out and separating the various sounds.

I thought it might be fun for us today to take a trip under water and listen to the sounds we might hear if we could actually visit the deep sea. [. . .]

⌣ 11 ⌣

[1951]

Jacket Notes for the RCA Victor Recording of Claude Debussy's "La Mer"/ National Symphony Orchestra Speech

Rachel Carson had no particular musical train-ing, but she did have a poet's understanding of the sea and a unique way of expressing her thoughts about it. After The Sea Around Us *appeared, a representative of RCA Victor records invited her to write the jacket notes for a new NBC Symphony recording of Claude Debussy's "La Mer," with Arturo Toscanini conducting. Her jacket notes do not explicate Debussy's music but suggest a different, no less lyrical, in-terpretation of the meaning and mystery of the sea and of the ancient world from which all life began.*

A short time later, Carson was invited to speak at a small benefit luncheon for the National Symphony Orchestra in Washington, D.C. With President Truman's wife, Bess, in atten-

dance, Carson commented briefly on how the sea influenced the music of Debussy, Rimsky-Korsakov, and Sibelius. Her remarks on the role of the arts in times of crisis reflect the tensions produced by the Korean War. In what would become a theme of her future public comments, Carson suggests that contemplation of the long history of the earth can bring comfort and reassurance to people in desperate times. This speech also contains Carson's first reference to the anxieties of living in the atomic age.

Jacket Notes for Debussy's "La Mer"

Claude Achille Debussy was born in St. Germain-en-Laye, France, in 1862. In boyhood years he seems to have been strongly attracted to the sea; this, and his father's hopes and ambitions for him pointed strongly toward a career in the Navy. Instead, "the chances of life" made him a musician. But Debussy, the composer, eventually returned to the dreams of his youth in one of his greatest compositions, *La Mer*, in which his intended and his chosen professions meet in brilliant synthesis.

So closely were Debussy's emotions attuned to the sea that he confessed himself almost overwhelmed and benumbed in its presence. He could not compose easily within sight or sound of it, but rather in some inland spot from which his recollections could return in tranquillity to the beauty and power and mystery of the sea. And certainly there was more than factual memory that came to him. There must have been also an intuitive perception of the mysterious inner nature of the sea, of truths which the science of the ocean, in its infancy in Debussy's time, had not yet discovered. We, who know some of these truths today, can discern them in this exquisitely beautiful evocation of the spirit of the sea.

Out of his "endless store of memories," Debussy has created a world of water and sky, crossed by the hurrying forms of waves and holding endless converse with the great winds that ceaselessly blow over the surface of the earth. It is a timeless, elemental world, in which the passage of the years and the centuries and the eons are lost in time itself — a world that might be of the Archeozoic Era or of the Twentieth Century.

The three movements of *La Mer* are titled: 1) From dawn till noon on the sea; 2) Play of the waves; 3) Dialogue of the wind and the sea. These titles might suggest that the composer was preoccupied with surface manifestations, and indeed, the music is full of the shimmering beauty of the face of the sea and the sparkle of sun on water. But as the surface of the sea itself is the creation and the expression of the unseen depths beneath it, so, underlying his musical recreation of the coming of dawn to the sea and of the wind-driven processions of the waves across the ocean, Debussy has suggested the mysterious and brooding spirit of the deep and hidden waters.

In the serene music of the first movement there is all the evanescent beauty of the first coming of light across the sea, the tenuous, pure airs of dawn moving over the water when the east turns grey and the black wave shapes come ashimmer with silver light. The face of the sea is mobile, sensitive, always changing. As the hours advance, changing lights and colors and the shifting shadows of the clouds move across its surface. More deliberate and subtle is the descent of dawn into deeper waters. Fathom by fathom the light steals down toward the threshold of the deep sea, a thousand feet or more below the surface. Only the noonday sun, with its long, straight rays, has power to penetrate to that transition zone between the surface waters and the eternal night of the abyss; so in these deep waters, the brief hour of dawn passes quickly into the hour of twilight, and the blue light fades away into the long night.

★ ★ ★

The sea is never at rest. The thin interface between air and water is exquisitely sensitive to the slightest disturbance. A drop of rain, a seabird coming down to alight on the water, a fish cutting the surface with its fin, set spreading ripples in motion. And always the winds, blowing over the face of the globe, are pushing up the water into the moving ridges of waves. The open sea is a playground of waves created by many different winds, rolling on diverse paths, intermingling, overtaking, passing, or sometimes engulfing one another. Born of wind and water, each young wave takes its place in the confused pattern of the open sea. Drawing energy from the winds that created them, the waves respond to the fury of the storm, trailing white streamers of foam, leaping up into steep, peaked shapes, crowding upon their fellows in a wild, abandoned play. In the wide immensity of the open sea, a wave knows no restraint; were it not for the intercepting masses of the continents it might roll on and on around the earth. But nearing shore, it feels the alien land beneath it. Against the drag of shoaling bottom its speed slackens. Within the surf zone it suddenly rears high, as though gathering strength against an unknown adversary. A white, foaming crest begins to form along its advancing front, and suddenly this shining creation of the open sea plunges forward and dissolves in thunder.

The third movement of *La Mer* introduces a sterner mood in this ancient dialogue of the wind and the waters. Hearing it, we think of the great wind belts where the westerly winds blow across thousands of miles of open sea and the most majestic of all waves march with them around the globe. Of such winds and such waves are born the terrible surf of Tierra del Fuego, or the violent seas that burst upon the shores of the Orkneys, when air and sea and land are blended in a thick obscurity of spray and leaping foam and beating waves.

The waves are the most eloquent of the sea's voices. In their wordless language they speak of the shrieking gales of the southern ocean, of the great anticyclonic winds sweeping around the Icelandic low, or they run directly ahead of an approaching storm, crying a warning. As they roll majestically in open ocean or as they break and surge at the edge of land, their voices are the voice of the sea.

What is this sea, and wherein lies its power so greatly to stir the minds of men? What is the mystery of it, intangible, yet inseparably its own? Perhaps part of the mystery resides in its hoary antiquity, for the sea is almost as old as earthly time. Its shadowy beginnings lie somewhere in that dim period when the earth was forming out of chaos, when deep basins were hollowed out of the cooling rocks and the rains began to fall from the thick cloud blanket that enveloped the earth. The rains poured upon the waiting basins, or falling upon the continents, drained away to become sea. And there began at once that slow erosion by which the continents are giving up their substance to the sea, by which the minerals are passing from earth to sea, and the sea is becoming ever more briny with the passing eons.

Or perhaps the spirit of the sea resides in the implacable, inexorable power by which it draws all things to it, by which it overwhelms and devours and destroys. The rivers run to it; the rains that rose from it return. For more than two billion years this sea has endured, changing yet seemingly changeless, while mountains have risen and been worn away, while islands have grown up from its floor, only to dissolve under the attack of rain and waves, and while the continents themselves have known the slow advance of engulfing seas, and again their slow retreat.

Or perhaps the mystery is the mystery of life itself — of life that began as a primordial bit of protoplasm adrift in the surface waters of the ancient seas. For hundreds of millions of years, all life was sea life, developing in prodigious abundance and variety, evolving into thousands of kinds of creatures, some of which finally crept out of the sea, some of which, after long eons of time, became men. But we as man carry the sea's salt in our blood, and the trace of our marine heritage in our bodies, and perhaps something akin to a racial memory of that dim past lies within us.

A sense of some of these things may come to one who makes a long ocean voyage — when day after day he watches the receding rim of the horizon ridged and furrowed by waves; or when he stands alone in darkness on the deck at night, in a world compounded only of water and sky, and feels the brooding presence of the sea about him. And surely the sense of these things was in Debussy's mind when he composed *La Mer*, capturing in immortal music the shining beauty, the awful power, and the eternal mystery of the sea.

National Symphony Orchestra Speech

[. . .] I believe quite sincerely that in these difficult times we need more than ever to keep alive those arts from which men derive inspiration and courage and consolation — in a word, strength of spirit. I believe this more strongly because of my own recent experiences — if I may again speak quite personally.

After my book was published I began to receive a great deal of mail. These letters are still coming. They are from people of all ages, and both sexes, and of all degrees of education.

They have made it clear that men and women in all walks of life are responding in a surprising way to what I have written about the ocean. They are finding in it something that is helping them face the problems of these difficult times.

That "something" is, I think, a new sense of perspective on human problems. When we contemplate the immense age of earth and sea, when we get in the frame of mind where we can speak easily of "millions" or "billions" of years, and when we remember the short time that human life has existed on earth, we begin to see that some of the worries and tribulations that concern us are very minor. We also gain some sense of confidence that the changes and the evolution of new ways of life are natural and on the whole desirable.

It has come to me very clearly through these wonderful letters that people everywhere are desperately eager for whatever will lift them out of themselves and allow them to believe in the future.

I am sure that such release from tension can come through the contemplation of the beauties and mysterious rhythms of the natural world.

But I am also sure that it is to be had through music in its reflection of the amazing creative genius of man.

We need the inspiration that comes from hearing great music. The symphony orchestras that present and interpret the music of the ages are not luxuries in this mechanized, this atomic age. They are, more than ever, necessities.

∿ 12 ∿

[1952]

Remarks at the Acceptance of the National Book Award for Nonfiction

In January, 1952, Carson learned she had won the prestigious National Book Award for nonfiction for The Sea Around Us. *At the New York award ceremony, where Carson was joined on the dais by James Jones, the fiction winner for* From Here to Eternity, *and poet Marianne Moore, critic John Mason Brown acknowledged that "Carson has atomized our egos and brought to each reader not only a new humility but a new sense of the inscrutable vastness and interrelation of forces beyond our knowledge or control."*

Carson used the occasion to comment on the isolation of science in America and on what she viewed as the artificial separation of science and literature as exclusive methods of investigating the world. Carson's early critique of the two cul-

146

tures mirrored that later made famous by the English scientist C. P. Snow in 1959.

Writing a book has surprising consequences, and the real education of the author perhaps begins on publication day. I, as the author, did not know how people would react to a book about the ocean. I am still finding out. When I planned my book, I knew only that a fascination for the sea and a compelling sense of its mystery had been part of my own life from earliest childhood. So I wrote what I knew about it and also what I thought and felt about it.

Many people have commented with surprise on the fact that a work of science should have a large popular sale. But this notion that "science" is something that belongs in a separate compartment of its own, apart from everyday life, is one that I should like to challenge. We live in a scientific age; yet we assume that knowledge of science is the prerogative of only a small number of human beings, isolated and priestlike in their laboratories. This is not true. The materials of science are the materials of life itself. Science is part of the reality of living; it is the what, the how, and the why of everything in our experience. It is impossible to understand man without understanding his environment and the forces that have molded him physically and mentally.

The aim of science is to discover and illuminate truth. And that, I take it, is the aim of literature, whether biography or history or fiction. It seems to me, then, that there can be no separate literature of science.

My own guiding purpose was to portray the subject of my sea profile with fidelity and understanding. All else was secondary. I did not stop to consider whether I was doing it scientifically or poetically; I was writing as the subject demanded.

The winds, the sea, and the moving tides are what they are. If there is wonder and beauty and majesty in them, science will discover these qualities. If they are not there, science cannot create them. If there is poetry in my book about the sea, it is not because I deliberately put it there, but because no one could write truthfully about the sea and leave out the poetry. [. . .]

We have looked first at man with his vanities and greed and his problems of a day or a year; and then only, and from this biased point of view, we have looked outward at the earth he has inhabited so briefly and at the universe in which our earth is so minute a part. Yet these are the great realities, and against them we see our human problems in a different perspective. Perhaps if we reversed the telescope and looked at man down these long vistas, we should find less time and inclination to plan for our own destruction.

◡∴ 13 ∴◡

[1952]

Design for Nature Writing

The John Burroughs medal awarded for excellence in nature writing was the one award Rachel Carson coveted. When she claimed the prize for The Sea Around Us *in a gala ceremony in New York in April 1952, she used the occasion to make some trenchant criticisms of the parochial attitudes of nature writers, chiding them for not trying hard enough to educate the public about the importance of natural science as a way of understanding the modern world. Carson was once again ahead of her time in suggesting that the public wanted more information about nature and natural history. She believed nature writers had a moral obligation to bring the wonders of the living world to the general public and urged them to accept that responsibility.*

In presenting me with the John Burroughs Medal you have welcomed me into an illustrious company, and you have given *The Sea Around Us* one of its most cherished honors. Any writer in the field of the natural sciences should feel a certain awe and even a sense of unreality in being linked during his or her own lifetime with the immortals in the field of nature writing. The tradition of John Burroughs, which you seek to keep alive through these awards, is a long and honorable one. It is a tradition that had its beginnings in even earlier writings. On the other side of the Atlantic it flowered most fully in the works of Richard Jefferies and W. H. Hudson; and in this country the pen of Thoreau — as that of John Burroughs himself — most truly represented the contemplative observer of the world about us. These four, I think, were the great masters. To those of us who have come later, there can scarcely be any greater honor than to be compared to one of them.

Yet if we are true to the spirit of John Burroughs, or of Jefferies or Hudson or Thoreau, we are not imitators of them but — as they themselves were — we are pioneers in new areas of thought and knowledge. If we are true to them, we are the creators of a new type of literature as representative of our own day as was their own.

I myself am convinced that there has never been a greater need than there is today for the reporter and interpreter of the natural world. Mankind has gone very far into an artificial world of his own creation. He has sought to insulate himself, in his cities of steel and concrete, from the realities of earth and water and the growing seed. Intoxicated with a sense of his own power, he seems to be going farther and farther into more experiments for the destruction of himself and his world.

There is certainly no single remedy for this condition and I am offering no panacea. But it seems reasonable to believe — and I do believe — that the more clearly we can focus our attention on the wonders and realities of the universe about us the less taste we shall have for the destruction of our race. Wonder and humility are wholesome emotions, and they do not exist side by side with a lust for destruction.

All of us here tonight are united by the strong bond of a common interest. In one way or another all of us have been touched by an awareness of the world of nature. No one present needs to be "sold" on this subject. But I should like to talk briefly about the non-naturalists and our attitude toward them — that large segment of the public that does not belong to the John Burroughs Association or to Audubon Societies and that really has very little knowledge of natural science. I am convinced that we have been far too ready to assume that these people are indifferent to the world we know to be full of wonder. If they are indifferent it is only because they have not been properly introduced to it — and perhaps that is in some measure our fault.

Since I am speaking of the John Burroughs Medal and what it means, perhaps I should confine my illustration to nature writing. I feel that we have too often written only for each other. We have assumed that what we had to say would interest only other naturalists. We have too often seemed to consider ourselves the last representatives of a dying tradition, writing for steadily dwindling audiences.

It is difficult to say these things without seeming to refer too directly to *The Sea Around Us*. Yet I feel they ought to be said, for in justice not only to ourselves but to the public we ought to develop a more confident and assured attitude toward the role and the value of nature literature. I am certain that what happened to *The Sea Around Us* could happen to many another book in the field of the natural sciences — and that it should happen.

Perhaps writers and publishers and magazine editors have all been at fault in taking, too often, a deprecating attitude which assumes in advance that a nature book will not have a wide audience, that it cannot possibly be a "commercial success."

This attitude is not only psychologically unsound; it is a mistaken and ill-founded one. The public is trying to show us how mistaken it is, if we will only listen. It proves our mistake when it fills Audubon Screen Tour showings with overflow audiences. It proves it when it buys Roger Peterson's bird guides by the many scores of thousands and goes afield with guide and binoculars. And if I may use a personal illustration, the letters that have come to me in the past nine months have taught me never again to underestimate the capacity of the general public to absorb the facts of science.

If these letters mean anything it is this: that there is an immense and unsatisfied thirst for understanding of the world about us, and every drop of information, every bit of fact that serves to free the reader's mind to roam the great spaces of the universe, is seized upon with almost pathetic eagerness.

I have learned from these letters, too, if I did not fully realize it before, that those who hunger for knowledge of their world are as varied as the passengers in a subway. The mail the other day brought letters from a

Catholic sister in a Tennessee school, a farmer in Saskatchewan, a British scientist, and a housewife. There have been hairdressers and fishermen and musicians and classical scholars and scientists. So many say, in one phrasing or another: "We have been troubled about the world, and had almost lost faith in man; it helps to think about the long history of the earth, and of how life came to be. When we think in terms of millions of years, we are not so impatient that our own problems be solved tomorrow."

These are the people who want to know about the world that is our chosen one. If we have ever regarded our interest in natural history as an escape from the realities of our modern world, let us now reverse this attitude. For the mysteries of living things, and the birth and death of continents and seas, are among the great realities.

The John Burroughs Medal is the only literary award that recognizes achievement in nature writing. In so doing, it may well be a force working toward a better civilization, by focusing attention on the wonders of a world known to so few, although it lies about us every day.

✌ 14 ✌

[1953]

Mr. Day's Dismissal

Albert M. Day was named director of the Fish and Wildlife Service in 1946, and under his direction the Service became the premier advocate for the conservation of the wildlife resources of the nation.

When the Republicans won the White House in 1952, they began to institute policies more beneficial to big business than to conservation. Shortly after Oregon businessman Douglas McKay was appointed Secretary of the Interior, Albert Day and other top professional staff of the department were dismissed and replaced by nonprofessional political appointees. Deeply disturbed by this trend and what it portended for the future of the environment, Carson took pen to paper to protest.

Her letter to the editor of the Washington Post

displays Carson's activist instincts, and her skill in raising the standard of political debate. Her letter was picked up by the Associated Press wire service, syndicated across the country, and later reprinted in Reader's Digest.

The dismissal of Mr. Albert M. Day as director of the Fish and Wildlife Service is the most recent of a series of events that should be deeply disturbing to every thoughtful citizen. The ominous pattern that is clearly being revealed is the elimination from the Government of career men of long experience and high professional competence and their replacement by political appointees. The firing of Mr. Marion Clawson, director of the Bureau of Land Management, is another example. There are widespread rumors that the head of the Park Service, who, like Mr. Day, has spent his entire professional life in the organization he heads, will also be replaced. These actions strongly suggest that the way is being cleared for a raid upon our natural resources that is without parallel within the present century.

The real wealth of the Nation lies in the resources of the earth — soil, water, forests, minerals, and wildlife. To utilize them for present needs while insuring their preservation for future generations requires a delicately balanced and continuing program, based on the most extensive research. Their administration is not properly, and cannot be, a matter of politics.

By long tradition, the agencies responsible for these resources have been directed by men of professional stature and experience, who have understood, respected, and been guided by the findings of their scientists. Mr. Day's career in wildlife conservation began 35 years ago, when, as a young biologist, he was appointed to the staff of the former Biological Survey, which later became part of the Fish and Wildlife Service. During the intervening years he rose through the ranks, occupying successively higher positions until he was appointed director in 1946. He achieved a reputation as an able and fair-minded administrator, with courage to stand firm against the minority groups who demanded that he relax wildlife conservation measures so that they might raid these public resources. Secretary McKay, whose own grasp of conservation problems is yet to be demonstrated, has now decreed that the Nation is to be deprived of these services.

These actions within the Interior Department fall into place beside the proposed giveaway of our offshore oil reserves and the threatened invasion of national parks, forests and other public lands.

For many years public-spirited citizens throughout the country have been working for the conservation of the natural resources, realizing their vital importance to the Nation. Apparently their hard-won progress is to be wiped out, as a politically minded Administration returns us to the dark ages of unrestrained exploitation and destruction.

It is one of the ironies of our time that, while concentrating on the defense of our country against enemies from without, we should be so heedless of those who would destroy it from within.

∿ 15 ∿

[1961]

Preface to the Second Edition of
The Sea Around Us

Working frantically to finish writing Silent
Spring *in 1961, Carson had not planned to take
time out to make changes in the new edition of*
The Sea Around Us. *But the publication of a
juvenile edition of her classic alerted her to certain
areas outdated by the rapid advances in oceanography in the decade since its publication. For the
new edition she chose not to alter the text itself,
but to amend it by adding new material as foot-
notes in an appendix, and to write a new preface.
Although Carson was then working in a field far
removed from oceanography, she kept up with
the latest research as an elected member of the
corporate board of the Woods Hole Marine Biological Laboratory and corresponded regularly
with marine scientists there.*

By 1961 Carson was deeply disturbed by the

evidence of humankind's despoiling reach even into those parts of the earth she had once considered inviolate. In her new preface, Carson outlines the ominous problem of dumping atomic waste at sea, and records her fear for the future if radioactive chemicals were to continue to contaminate the ocean's food chains and ecosystems.

The sea has always challenged the minds and imagination of men and even today it remains the last great frontier of Earth. It is a realm so vast and so difficult of access that with all our efforts we have explored only a small fraction of its area. Not even the mighty technological developments of this, the Atomic Age, have greatly changed this situation. The awakening of active interest in the exploration of the sea came during the Second World War, when it became clear that our knowledge of the ocean was dangerously inadequate. We had only the most rudimentary notions of the geography of that undersea world over which our ships sailed and through which submarines moved. We knew even less about the dynamics of the sea in motion, although the ability to predict the actions of tides and currents and waves might easily determine the success or failure of military undertakings. The practical need having been so clearly established, the governments of the United States and of other leading sea powers began to devote increasing effort to the scientific study of the sea. Instruments and equipment, most of which had been born of urgent necessity, gave oceanographers the means of tracing the contours of the ocean bottom, of studying the movements of deep waters, and even of sampling the sea floor itself.

These vastly accelerated studies soon began to show that many of the old conceptions of the sea were faulty, and by the mid-point of the century a new picture had begun to emerge. But it was still like a huge canvas on which the artist has indicated the general scheme of his grand design but on which large blank areas await the clarifying touch of his brush.

This was the state of our knowledge of the ocean world when *The Sea Around Us* was written in 1951. Since that time the filling in of many of the blank areas has proceeded and new discoveries have been made. In this second edition of the book I have described the most important of the new findings in a series of notes which will be found in the Appendix. These notes are keyed to appropriate passages in the original text by reference numbers. For example, after the discussion of the Arctic Ocean ending on page 65, the reader may go on to learn of recent discoveries in this area by turning to note 10 in the Appendix.

The 1950's have comprised an exciting decade in the science of the sea. During this period a manned vehicle has descended to the deepest hole in the ocean floor. During the fifties, also, the crossing of the entire Arctic basin was accomplished by submarines traveling under the ice. Many new features of the unseen floor of the sea have been described, including new mountain ranges that now appear to be linked with others to form the longest and mightiest mountains of the earth — a continuous chain encircling the globe. Deep, hidden rivers in the sea, subsurface currents with the volume of a thousand Mississippis, have been found. During the International Geophysical Year, 60 ships from 40 nations, as well as hundreds of stations on islands and seacoasts, cooperated in an enormously fruitful study of the sea.

Yet the present achievements, exciting though they are, must be considered only a beginning to what is yet to be achieved by probing the vast depths of water that cover most of the surface of the earth. In 1959 a group of distinguished scientists comprising the Committee on Oceanography of the

National Academy of Sciences declared that "Man's knowledge of the oceans is meager indeed compared with their importance to him." The Committee recommended at least a doubling of basic research on the sea by the United States in the 1960's; anything less would, in its opinion, "jeopardize the position of oceanography in the United States" compared with other nations and "place us at a disadvantage in the future use of the relief of the lower slopes."

One of the most fascinating of the projects now planned for the future is an attempt to explore the interior of the earth by drilling a hole three or four miles deep in the bottom of the sea. This project, which is sponsored by the National Academy of Sciences, is designed to penetrate farther than instruments have ever before reached, to the boundary between the earth's crust and its mantle. This boundary is known to geologists as the Mohorovicic discontinuity (or more familiarly as the Moho) because it was discovered by a Yugoslavian of that name in 1912. The Moho is the point at which earthquake waves show a marked change in velocity, indicating a transition from one kind of material to something quite different. It lies much deeper under the continents than under the oceans, so, in spite of the obvious difficulties of drilling in deep water, an ocean site offers most promise. Above the Moho lies the crust of the earth, composed of relatively light rocks, below it the mantle, a layer some 1800 miles thick enclosing the hot core of the earth. The composition of the crust is not fully known and the nature of the mantle can be deduced only by the most indirect methods. To penetrate these regions and bring back actual samples would therefore be an enormous step forward in understanding the nature of our earth, and would even advance our knowledge of the universe, since the deep structure of the earth may be assumed to be like that of other planets.

As we learn more about the sea through the combined studies of many specialists a new concept that is gradually taking form will almost certainly be strengthened. Even a decade or so ago it was the fashion to speak of the abyss as a place of eternal calm, its black recesses undisturbed by any movement of water more active than a slowly creeping current, a place isolated from the surface and from the very different world of the shallow sea. This picture is rapidly being replaced by one that shows the deep sea as a place of movement and change, an idea that is far more exciting and that possesses deep significance for some of the most pressing problems of our time.

In the new and more dynamic concept, the floor of the deep sea is shaped by racing turbidity currents or mud flows that pour down the slopes of the ocean basins at high speed; it is visited by submarine landslides and stirred by internal tides. The crests and ridges of some of the undersea mountains are swept bare of sediments by currents whose action, in the words of geologist Bruce Heezen, is comparable to "snow avalanches in the Alps (which) sweep down and smother the relief of the lower slopes."

Far from being isolated from the continents and the shallow seas that surround them, the abyssal plains are now known to receive sediments from the margins of the continents. The effect of the turbidity currents, over the vast stretches of geologic time, is to fill the trenches and the hollows of the abyssal floor with sediment. This concept helps us understand certain hitherto puzzling occurrences. Why, for example, have deposits of sand — surely a product of coastal erosion and the grinding of surf — appeared on the mid-ocean floor? Why have sediments at the mouths of submarine canyons, where they communicate with the abyss, been found to contain such reminders of the land as bits of wood and leaves, and why are there sands containing nuts, twigs, and the bark of trees even farther out on the plains of the abyss? In the powerful downrush of sediment-laden currents, triggered by storms or floods or earthquakes, we now have a mechanism that accounts for these once mysterious facts.

Although the beginnings of our present concept of a dynamic sea go back perhaps several decades, it is only the superb instruments of the past ten years that have allowed us to glimpse the hidden movements of ocean waters. Now we suspect that all those dark regions between the surface and the bottom are stirred by currents. Even such mighty surface currents as the Gulf Stream are not quite what we supposed them to be. Instead of a broad and steadily flowing river of water, the Gulf Stream is now found to consist of narrow racing tongues of warm water that curl back in swirls and eddies. And below the surface currents are others unlike them, running at their own speeds, in their own direction, with their own volume. And below these are still others. Photographs of the sea bottom taken at great depths formerly supposed to be eternally still show ripple marks, a sign that moving waters are sorting over sediments and carrying away the finer particles. Strong currents have denuded the crest of much of the vast range of undersea mountains known as the Atlantic Ridge, and every one of the sea mounts that has been photographed reveals the work of deep currents in ripple marks and scour marks.

Other photographs give fresh evidence of life at great depths. Tracks and trails cross the sea floor and the bottom is studded with small cones built by unknown forms of life or with holes inhabited by small burrowers. The Danish research vessel *Galathea* brought up living animals in dredges operated at great depths, where only recently it was supposed life would be too scanty to permit such sampling.

These findings of the dynamic nature of the sea are not academic; they are not merely dramatic details of a story that has interest but no application. They have a direct and immediate bearing on what has become a major problem of our time.

Although man's record as a steward of the natural resources of the earth has been a discouraging one, there has long been a certain comfort in the belief that the sea, at least, was inviolate, beyond man's ability to change and to despoil. But this belief, unfortunately, has proved to be naïve. In unlocking the secrets of the atom, modern man has found himself confronted with a frightening problem — what to do with the most dangerous materials that have ever existed in all the earth's history, the by-products of atomic fission. The stark problem that faces him is whether he can dispose of these lethal substances without rendering the earth uninhabitable.

No account of the sea today is complete unless it takes note of this ominous problem. By its very vastness and its seeming remoteness, the sea has invited the attention of those who have the problem of disposal, and with very little discussion and almost no public notice, at least until the late fifties, the sea has been selected as a "natural" burying place for the contaminated rubbish and other "low-level wastes" of the Atomic Age.* These wastes are placed in barrels lined with concrete and hauled out to sea, where they are dumped overboard at previously designated sites. Some have been taken out 100 miles or more; recently sites only 20 miles offshore

---

* Ed.: Carson is referring to the practice of dumping radioactive waste at sea which the U.S. and other nuclear powers had been doing since 1946. The U.S. stopped dumping atomic waste in 1970 at the recommendation of the U.S. Council on Environmental Quality and signed and ratified the London Dumping Convention in 1972. Russia was found in violation of high-level dumping in the early 1990s, and England, France, and Russia still continue to dump low-level waste in the ocean.

have been suggested. In theory the containers are deposited at depths of about 1000 fathoms, but in practice they have at times been placed in much shallower waters. Supposedly the containers have a life of at least 10 years, after which whatever radioactive materials remain will be released to the sea. But again this is only in theory, and a representative of the Atomic Energy Commission, which either dumps the wastes or licenses others to do so, has publicly conceded that the containers are unlikely to maintain "their integrity" while sinking to the bottom. Indeed, in tests conducted in California, some have been found to rupture under pressure at only a few hundred fathoms.

But it is only a matter of time until the contents of all such containers already deposited at sea will be free in the ocean waters, along with those yet to come as the applications of atomic science expand. To the packaged wastes so deposited there is now added the contaminated runoff from rivers that are serving as dumping grounds for atomic wastes, and the fallout from the testing of bombs, the greater part of which comes to rest on the vast surface of the sea.

The whole practice, despite protestations of safety by the regulatory agency, rests on the most insecure basis of fact. Oceanographers say they can make "only vague estimates" of the fate of radioactive elements introduced into the deep ocean. They declare that years of intensive study will be needed to provide understanding of what happens when such wastes are deposited in estuaries and coastal waters. As we have seen, all recent knowledge points to far greater activity at all levels of the sea than had ever been guessed at. The deep turbulence, the horizontal movements of vast rivers of ocean water streaming one above another in varying directions, the upwelling of water from the depths carrying with it minerals from the bottom, and the opposite downward sinking of great masses of surface water, all result in a gigantic mixing process that in time will bring about universal distribution of the radioactive contaminants.

And yet the actual transport of radioactive elements by the sea itself is only part of the problem. The concentration and distribution of radioisotopes by marine life may possibly have even greater importance from the standpoint of human hazard. It is known that plants and animals of the sea pick up and concentrate radiochemicals, but only vague information now exists as to details of the process. The minute life of the sea depends for its existence on the minerals in the water. If the normal supply of these is low, the organisms will utilize instead the radioisotope of the needed element if it is present, sometimes concentrating it as much as a million times beyond its abundance in sea water. What happens then to the careful calculation of a "maximum permissible level"? For the tiny organisms are eaten by larger ones and so on up the food chain to man. By such a process tuna over an area of a million square miles surrounding the Bikini bomb test developed a degree of radioactivity enormously higher than that of the sea water.

By their movements and migrations, marine creatures further upset the convenient theory that radioactive wastes remain in the area where they are deposited. The smaller organisms regularly make extensive vertical movements upward toward the surface of the sea at night, downward to great depths by day. And with them goes whatever radioactivity may be adhering to them or may have become incorporated into their bodies. The larger fauna, like fishes, seals, and whales, may migrate over enormous distances, again aiding in spreading and distributing the radioactive elements deposited at sea.

The problem, then, is far more complex and far more hazardous than has been admitted. Even in the comparatively short time since disposal began, research has shown that some of the assumptions on which it was based were dangerously inaccurate. The truth is that disposal has proceeded far more rapidly than our knowledge justifies. To dispose first and investigate later is an invitation to disaster, for once radioactive elements have been deposited at sea they are irretrievable. The mistakes that are made now are made for all time.

It is a curious situation that the sea, from which life first arose, should now be threatened by the activities of one form of that life. But the sea, though changed in a sinister way, will continue to exist; the threat is rather to life itself.

Part Three

Part Three begins with Carson's most successful magazine article, "Our Ever-Changing Shore" from Holiday, *and ends with a television script on clouds. Carson had long understood that the same physical forces were at work in the air and on the sea, a similarity that fascinated her.*

After leaving the Fish and Wildlife Service in 1952 Rachel Carson bought a cottage on Southport Island, Maine, near Boothbay Harbor, where with her own tide pools and rocky beach she spent the middle years of the decade researching and writing The Edge of the Sea, *the final volume in her trilogy on the sea. As with her previous book, it was serialized first in the* New Yorker *and then appeared on the* New York Times *best-seller list.*

Field research on the Atlantic seashores provided Carson with some of her most creative moments, several of which are reflected in the selections included here from her field notebooks and from her letters to friends. Carson used those field experiences in public speeches during this pe-

riod to reflect on the primacy of ecological relationships and the value of beauty in the modern world.

Concerned about the dwindling number of America's seashores, Carson advocated that some must be set aside from human activity. She had been an early advocate of wilderness preservation, but she also dreamed of saving a small tract of land she called "Lost Woods" on Southport Island. Although this ambition was never achieved, Carson left a written legacy of her advocacy for this and for the ecological importance of life in the marginal world.

∿ 16 ∿

[1958]

Our Ever-Changing Shore

The editors of the new Holiday *magazine planned a special issue for the summer of 1958 devoted to "Nature's America," and asked Rachel Carson to contribute a short article on the nation's seashores. Although Carson's attention had already turned to the problems of pesticide misuse, she accepted because, as she told her literary agent, the article provided the opportunity to "get in a few licks about how few seashores remain." Carson hoped that if she effectively communicated the threat to the nation's wild seashores, some might be saved.*

Carson drew her examples from descriptions of coastlines she had explored and written about in letters to friends, and from observations she had made in her field notes. The resulting article combined acute description with an intimate feeling of being a fellow traveler with Carson to the shores she knew and loved. Her plea for the preservation of the nation's seashores remains one of the most eloquent in contemporary nature writing.

Along mile after mile of coastline, the land presents a changing face to the sea. Now it is a sheer rock cliff; now a smooth beach; now the frayed edge of a mangrove swamp, dark and full of mystery. Each is the seacoast, yet each is itself, like no other in time or place. In every outthrust headland, in every curving beach, in every grain of sand there is a story of the earth.

This coastline plays endless variations on the basic theme of sea and land. On the coastal rocks of northern New England the sea is an immediate presence, compelling, impossible to ignore. Its tides rise and fall on their appointed schedule, draining coves and refilling them, lifting boats or dropping away to leave them stranded. On the broad beaches of the South the feeling is different. As you stand at the edge of the dunes, when the tide is out, the ocean seems far away. Under the push of a rising tide it advances a little, reducing the width of the buffer strip of sand. Storms bring it still farther in. But compared with its overwhelming presence on Northern shores it seems remote, a shining immensity related to far horizons. The sound of the waves on such a day, when the heated air shimmers above the sand and the sky is without clouds, is a muted whisper. In this quiet there is a tentativeness that suggests that something is about to happen. And indeed we may be sure the present stand of the sea here is only temporary, for many times in the past million years or so it has risen and flowed across all of the coastal plain, paused for perhaps a few thousand years, and returned again to its basin.

For the shore is always changing, and today's sand beach may become the sheer rock coast of a distant tomorrow. This is precisely what happened in northern New England, where, only a few thousands of years ago, the earth's crust sank and the sea came in, covering the beaches and the plain, running up the river valleys and rising about the hills. So, on the young Maine coast today, evergreen forests meet the granite threshold of the sea.

Everywhere the wind and the sea have shaped the coast, sculpturing it into forms that are often beautiful, sometimes bizarre. Along the Oregon coast the rocky cliffs and headlands speak of the age-long battle with the sea. Here and there a lonely tower of rock rises offshore, one of the formations known as stacks or needles. Each began as a narrow headland jutting out from the main body of coastal rock. Then a weak spot in its connection with the mainland was battered through.

Here and there the assaults of surf have blasted out caves in the sea cliffs. Anemone Cave in Acadia National Park is one. In the famous Sea Lion Caves on the central Oregon coast several hundred sea lions gather each autumn, living in the tumultuous surge of the surf, mingling their roars with the sound of the sea, still working to break through the roof of the cave.

Back from the surf line, the winds have piled up majestic dunes here and there. At Kitty Hawk in North Carolina perhaps the highest dunes of the American coast rise abruptly from the sea. I have stood on the summit of one of these dunes on a windy day when all the crest appeared to be smoking, and the winds seemed bent on destroying the very dunes they had created. Clouds and streamers of sand grains were seized by the strong flow of air and carried away. Far below, in the surf line, I could see the source of the dune sand, where the waves are forever cutting and grinding and polishing the fragments of rock and shell that compose the coastal sands.

The curving slopes, the gullies, the ridged surfaces of the dunes all carry the impress of the sea winds. So, in many places, do living things. The westerly winds that sweep across thousands of miles of open ocean at times pile up on our northern Pacific shores the heaviest surf of the whole Western Hemisphere. They are also the sculptors of the famous Monterey cypresses, the branches of which stream landward as though straining to escape the sea, though rooted near it. Actually the cutting edge that prunes such coastal vegetation is the sea salt with which the wind is armed, for the salt kills the growing buds on the exposed side.

The shore means many things to many people. Of its varied moods the one usually considered typical is not so at all. The true spirit of the sea does not reside in the gentle surf that laps a sun-drenched bathing beach on a summer day. Instead, it is on a lonely shore at dawn or twilight, or in storm or midnight darkness that we sense a mysterious something we recognize as the reality of the sea. For the ocean has nothing to do with humanity. It is supremely unaware of man, and when we carry too many of the trappings of human existence with us to the threshold of the sea world our ears are dulled and we do not hear the accents of sublimity in which it speaks.

Sometimes the shore speaks of the earth and its own creation; sometimes it speaks of life. If we are lucky in choosing our time and place, we may witness a spectacle that echoes vast and elemental things. On a summer night when the moon is full, the sea and the swelling tide and creatures of the ancient shore conspire to work primeval magic on many of the beaches from Maine to Florida. On such a night the horseshoe crabs move in, just as they did under a Paleozoic moon — just as they have been doing through all the hundreds of millions of years since then — coming out of the sea to dig their nests in the wet sand and deposit their spawn.

As the tide nears its flood dark shapes appear in the surf line. They gleam with the wetness of the sea as the moon shines on the curves of their massive shells. The first to arrive linger in the foaming water below the advancing front of the tide. These are the waiting males. At last other forms emerge out of the darkness offshore, swimming easily in the deeper water but crawling awkwardly and hesitantly as the sea shallows beneath them. They make their way to the beach through the crowd of jostling males. In thinning water each female digs her nest and sheds her burden of eggs, hundreds of tiny balls of potential life. An attending male fertilizes them. Then the pair moves on, leaving the eggs to the sea, which gently stirs them and packs the sand about them, grain by grain.

Not all of the high tides of the next moon cycle will reach this spot, for the water movements vary in strength and are weakest of all at the moon's quarters. A month after the egg laying the embryos will be ready for life; then the high tides of another full moon will wash away the sand of the nest. The turbulence of the rising tide will cause the egg membranes to split, releasing the young crabs to a life of their own over the shallow shores of bays and sounds.

But how do the parent crabs foresee these events? What is there in this primitive, lumbering creature that tells it that the moon is full and the tides are running high? And what tells it that the security of its eggs will be enhanced if the nests are dug and the eggs deposited on these stronger tides?

Tonight, in this setting of full moon and pressing tide, the shore speaks of life in a mysterious and magical way. Here is the sea and the land's edge. Here is a creature that has known such seas and shores for eons of time, while the stream of evolution swept on, leaving it almost untouched since the days of the trilobites. The horseshoe crabs in their being obliterate the barrier of time. Our thoughts become uncertain; is it really today? or is it a million — or a hundred million years ago?

Or sometimes when the place and mood are right, and time is of no account, it is the early sea itself that we glimpse. I remember feeling, once, that I had actually sensed what the young earth was like. We had come down to the sea through spruce woods — woods that were dim with drifting mists and the first light of day. As we passed beyond the last line of trees onto the rocks of the shore a curtain of fog dropped silently but instantly behind us, shutting out all sights and sounds of the land. Suddenly our world was only the dripping rocks and the gray sea that occasionally exploded in a muted roar. These, and the gray mists — nothing more. For all we could tell the time might have been Paleozoic, when the world was in very fact only rocks and sea.

We stood quietly, speaking few words. There was nothing, really, for human words to say in the presence of something so vast, mysterious and immensely powerful. Perhaps only in music of deep inspiration and grandeur could the message of that morning be translated by the human spirit as in the opening bars of Beethoven's Ninth Symphony — music that echoes across vast distances and down long corridors of time, bringing the sense of what was and of what is to come — music of swelling power that swirls and explodes even as the sea surged against the rocks below us.

But that morning all that was worth saying was being said by the sea. It is only in wild and solitary places that it speaks so clearly. Another such place I like to remember is that wilderness of beach and high dunes where Cape Cod, after its thirty-mile thrust into the Atlantic, bends back toward the mainland. Over thousands of years the sea and the wind have worked together to build this world out of sand. The wide beach is serene, like the ocean that stretches away to a far-off horizon. Offshore the dangerous shoals of Peaked Hill Bars lie just beneath the surface, holding within themselves the remains of many ships. Behind the beach the dunes begin to rise, moving inland like a vast sea of sand waves caught in a moment of immobility as they sweep over the land.

The dunes are a place of silence, to which even the sound of the sea comes as a distant whisper; a place where, if you listen closely, you can hear the hissing of the ever mobile sand grains that leap and slide in every breath of wind, or the dry swish of the beach grass writing, writing its endless symbols in the sand.

Few people come out through that solitude of dune and sky into the vaster solitude of beach and sea. A bird could fly from the highway to the beach in a matter of minutes, its shadow gliding easily and swiftly up one great desert ridge and down another. But such easy passage is not for the human traveler, who must make his slow way on foot. The thin line of his footprints, toiling up slopes and plunging down into valleys, is soon erased by the shifting, sliding sands. So indifferent are these dunes to man, so quickly do they obliterate the signs of his presence, that they might never have known him at all.

I remember my own first visit to the beach at Peaked Hill Bars. From the highway a sandy track led off through thickets of pine. The horizon lay high on the crest of a near dune. Soon the track was lost, the trees thinned out, the world was all sand and sky.

From the crest of the first hill I hoped for a view of the sea. Instead there was another hill, across a wide valley. Everything in this dune world spoke of the forces that had created it, of the wind that had shifted and molded the materials it received from the sea, here throwing the surface of a dune into firm ridges, there smoothing it into swelling curves. At last I came to a break in the seaward line of dunes and saw before me the beach and the sea.

On the shore below me there was at first no sign of any living thing. Then perhaps half a mile down the beach I saw a party of gulls resting near the water's edge. They were silent and intent, facing the wind. Whatever communion they had at that moment was with the sea rather than with each other. They seemed almost to have forgotten their own kind and the ways of gulls. When once a white, feathered form drifted down from the dunes and dropped to the sand beside them none of the group challenged him. I approached them slowly. Each time I crossed that invisible line beyond which no human trespasser might come; the gulls rose in a silent flock and moved to a more distant part of the sands. Everything in that scene caused me to feel apart, remembering that the relation of birds to the sea is rooted in millions of years, that man came but yesterday.

There have been other shores where time stood still. On Buzzards Bay there is a beach studded with rocks left by the glaciers. Barnacles grow on them now, and a curtain of rockweeds drapes them below the tide line. The bay shore of mud and sand is crossed by the winding trails of many periwinkles. On the beach at every high tide are cast the shells and empty husks of all that live offshore: the gold and silver shells of the rock oysters or jingles, the curious little half decks or slipper shells, the brown, fernlike remains of *Bugula*, the moss animal, the bones of fishes and the egg strings of whelks.

Behind the beach is a narrow rim of low dunes, then a wide salt marsh. This marsh, when I visited it on an evening toward the end of summer, had filled with shore birds since the previous night; their voices were a faint, continuous twittering. Green herons fished along the creek banks, creeping along at the edge of the tall grasses, placing one foot at a time with infinite care, then with a quick forward lunge attempting to seize some small fish or other prey. Farther back in the marsh a score of night herons stood motionless. From the bordering woods across the marsh a mother deer and her two fawns came down to drink silently, then melted back into their forest world.

The salt marsh that evening was like a calm, green sea — only a little calmer, a little greener than the wide sheet of the bay on the other side of the dunes. The same breeze that rippled the surface of the bay set the tips of the marsh grasses to swaying in long undulations. Within its depths the marsh concealed the lurking bittern, the foraging heron, the meadow mouse running down long trails overarched by grass stems, even as the watery sea concealed the lurking squids and fishes and their prey. Like the foam on the beach when the wind had whipped the surface waters into a light froth, the even more delicate foam of the sea lavender flecked the barrier of dunes and ran to the edge of the marsh. Already the fiery red of the glasswort or marsh samphire flickered over the higher ground of the marsh, while offshore mysterious lights flared in the waters of the bay at night. These were signs of approaching autumn, which may be found at the sea's edge before even the first leaf shows a splash of red or yellow.

The sea's phosphorescence is never so striking alongshore as in late summer. Then some of the chief light producers of the water world have their fall gatherings in bays and coves. Just where and when their constellations will form no one can predict. And the identity of these wheeling stars of the night sea varies. Usually the tiny glittering sparks are exceedingly minute, one-celled creatures, called dinoflagellates. Larger forms, flaring with a ghostly blue-white phosphorescence, may be comb jellies, crystal-clear and about the size of a small plum.

On beach and dune and over the flat vistas of salt marsh, too, the advancing seasons cast their shadows; the time of change is at hand. Mornings, a light mist lies over the marshes and rises from the creeks. The nights begin to hint of frost; the stars take on a wintry sparkle; Orion and his dogs hunt in the sky. It is a time, too, of color — red of berries in the dune thickets, rich yellow of the goldenrod, purple and lacy white of the wild asters in the fields. In the dunes and on the ocean beach the colors are softer, more subtle. There may be a curious purple shading over the sand. It shifts with the wind, piles up in little ridges of deeper color like the ripple marks of waves. When first I saw this sand on the northern Massachusetts coast, I wondered what it was. According to local belief the purple color comes from some seaweed, left on the shore, dried, and reduced to a thin film of powder. Years later I found the answer. I discovered drifts of the same purple color amid the coarse sand of my own shore in Maine — sand largely made up of broken shell and rock, fragments of sea-urchin spines, opercula of snails. I brought some of the purple sand to the house. When I put a pinch of it under the microscope I knew at once that this came from no plant — what I saw was an array of gems, clear as crystal, returning a lovely amethyst light to my eyes. It was pure garnet.

The sand grains scattered on the stage of my microscope spoke in their own way of the timeless, unhurried spirit of earth and sea. They were the end product of a process that began eons ago deep inside the earth, continued when the buried mineral was brought at last to the surface, and went on through millenniums of time and, it may be, through thousands of miles of land and sea until, tiny, exquisite gems of purest color, they came temporarily to rest at the foot of a glacier-scarred rock.

Perhaps something of the strength and serenity and endurance of the sea — of this spirit beyond time and place — transfers itself to us of the land world as we confront its vast and lonely expanse from the shore, our last outpost.

The shore might seem beyond the power of man to change, to corrupt. But this is not so. Unhappily, some of the places of which I have written no longer remain wild and unspoiled. Instead, they have been tainted by the sordid transformation of "development" — cluttered with amusement concessions, refreshment stands, fishing shacks — all the untidy litter of what passes under the name of civilization. And so noisy are these attributes of man that the sea cannot be heard. On all coasts it is the same. The wild seacoast is vanishing.

Five thousand miles of true ocean beach may seem inexhaustible wealth, but it is not. The National Park Service has recently published a survey of the remaining undeveloped areas on the Atlantic and Gulf coasts. (The results of a Pacific survey are yet to be released.) It described the situation it discovered as "foreboding," for "almost every attractive seashore area from Maine to Mexico that is accessible by road has been acquired for development purposes, or is being considered for its development possibilities. The seashore is rapidly vanishing from public use."

The Service asked that public-minded citizens and local, State and Federal Governments take the necessary steps, before it is too late, "to preserve this priceless heritage." Of the open shoreline of the Atlantic and Gulf coasts only 6 ½ percent is owned by the states or nation. The Park Service urged that at least 15 per cent of the general shoreline of our east coast should be publicly owned. This means acquiring an additional 320 miles at once. This must be done if we are to insure that we ourselves, and generations to follow, may know what the shore is like, may read the meaning and message of this strip between land and sea.

In its effort to awaken the public to the threatened loss of all natural seashore, the Park Service is recalling a recommendation made following a survey in 1935. Then, just a human generation ago (a mere second in earth history) the situation was very different. At that time the Park Service urged that 12 major strips, totaling 437 miles of beach, be preserved for public use. Only one of these was acquired. All the rest of these strips, except one, have since gone into private or commercial development.

One of the areas then recommended could have been bought at that time for $9000 a mile. Now, thanks to the post–World War II boom in seashore property, its price tag is $110,000 a mile.

To convert some of the remaining wild areas into State and National parks, however, is only part of the answer. Even public parks are not what nature created over the eons of time, working with wind and wave and sand. Somewhere we should know what was nature's way; we should know what the earth would have been had not man interfered. And so, besides public parks for recreation, we should set aside some wilderness areas of seashore where the relations of sea and wind and shore — of living things and their physical world — remain as they have been over the long vistas of time in which man did not exist. For there remains, in this space-age universe, the possibility that man's way is not always best.

᭙ 17 ᭙

[1950–1952]

Four Fragments from
Carson's Field Notebooks

The Edge of the Sea *was conceived as a guide to the shore life of the Atlantic coast. Beginning in 1950, Carson traveled from Maine to Florida studying tidal ecology. These fragments from Carson's field notebooks were written during several research trips to the remote beaches of the Carolinas and Georgia.*

Carson's field observations were never so narrow or self-absorbing that she missed the wider angle of vision or failed to relate her specific environment to the larger evolution of life. She was always an immediate participant intimately in touch with the life of her fellow creatures. Her notebooks testify to her compassion, her capacity for wonder, and her humility in the face of creation.

Saturday

Hiked north on beach. Very windy, a quick shower or two, much froth. Saw a little one-legged sanderling hopping along hunting food. Without my glasses I couldn't be sure whether the injured leg was cut off or drawn up under the body, but it was completely useless. Still he ran and probed, not venturing as near the surf as they usually do. When I came near he wheeled out over the water, his sharp "pit, pil" quickly lost in the sound of the waves. I thought of the long miles of travel ahead of him and wondered how long he would last. As I came back down the beach I saw him again still hopping along bravely.

A very few ghost crabs were out, but scuttled back quickly into their holes. I sat down on a box to wait for one to come out, feeling like a cat watching a mouse hole, but soon it began to rain and I moved on.

Saw tracks of a shore bird — probably a sanderling, and followed them a little, then they turned toward the water and were soon obliterated by the sea. How much it washes away, and makes as though it had never been. Time itself is like the sea, containing all that came before us, sooner or later sweeping us away on its flood and washing over and obliterating the traces of our presence, as the sea this morning erased the footprints of the bird.

On the way back I met the little one-legged sanderling again, the one I had seen Saturday afternoon. Remembering how the legs of the normal ones twinkle as they dash up and down the beach, it was amazing to see how fast this little fellow got about just hopping, hopping on his good right leg. This time I could see that his left leg is only a short stump less than an inch long. I wondered if some animal, maybe a fox, had caught it in the Arctic, or whether it had gotten into a trap. Their way of feeding being what it is, one would say he would have been eliminated before this as "unfit" — yet he must be even tougher than his two-legged comrades. That last word is a misnomer, he had no companions — either time I saw him — just hunting alone, he would hop, hop, hop, toward the surf; probing and jabbing busily with opened bill, turn and hop away from the advancing foam. Only twice did I see him have to take to his wings to escape a wetting. It made my heart ache to think how tired his little leg must be, but his whole manner suggested a cheerfulness of spirit and a gameness which must mean that the God of fallen sparrows has not forgotten him.

Little Dog

The next day, out on these same flats in early morning for low tide, I saw *Callianassa* scooting around in pools left in depressions — thanks to the help of a little dog. I first saw him away out on the flats, apparently by himself, and I thought he was chasing birds. There were the usual willets and a snowy egret and they'd get up and move on when he approached. But he was interested in the shallow pools of water and would wade in and go trotting around, his stumpy tail wagging constantly. I first wondered if he was noticing the little glittering reflections that were dancing all over the bottom, for the breeze kept little ripples stirring and the sun was very bright. When I came back down the beach later, he was still out, trotting around in the same pool. Everyone had gone in; the tide had turned, and I was worried for fear he might be cut off — he was very far out — get bewildered about where the shore was, and drown. So I decided to go out after him. He just wouldn't be distracted, but went on trotting in circles. Then I saw the darting of the little, almost transparent forms of shrimp and knew what was attracting him. In the end I had to pick him up and carry him a little distance; then he scampered ahead to another pool and resumed his shrimp hunting, but since that was near the upper beach I didn't worry. [. . .]

There are a fair number of *Diapatra* tubes [plumed worms] out on these flats. In some of the depressions (winding ones almost like creek beds) that seem always to have water even when tide is out — you see many tracks winding back and forth. When you can see where one ends (and sometimes there will be movement evident) you dig down and find a live moon snail moving along.

I think *Callianassa* holes differ in appearance according to the kind of bottom: where it is sandy you get the little excreted pellets looking like chocolate, and scattered closely around mouth of hole but not much mounded up. Where it is muddy, you seem to get elevated chimneys. The mud is sort of coiled, as though squeezed out of pastry tube with fancy fillips. [. . .] Some are flattened — others go up to a peak. On digging, I could get canal going down into sand, but could never get shrimp.

While digging, found empty *Cystoidean* tube.

Out here, I also watched sand dollars burying themselves in sand. You see a broad track, explore end of it carefully with fingers and find dollar. If you dig one out, it will immediately start to disappear into the sand — there is a considerable current stirred up all around edge, and body starts to dip under at forward edge. It takes it only a couple of minutes to disappear under sand again.

Saint Simon Island, Georgia (1952)

On the beach in front of the Coast Guard station, and from there north to the Inlet, an immense stretch of sand is exposed at low tide. One can walk out probably half a mile, almost dry-shod, it being necessary, here and there, to wade through just enough depth of water to wet one's shoes. The upper beach — i.e. from high tide line down perhaps several hundred feet — is smooth sand, but farther down there must be a mixture of mud or clay which gives a firmer consistency. This part of the beach, when the tide is out, is always deeply grooved with ripple marks — a pattern of wavelets sculptured and preserved for the tidal interval in this curiously firm substance. [. . .]

On the evening of April 17 I had a wonderful hour on these flats from 6:30–7:30, coming in almost at dark. Low tide was about 8:15, so the water still had almost an hour to ebb, but really incredible expanses of sand were exposed. Away out there, so far from the buildings on shore, it was nice to think that this wide tidal area belonged to the sea and couldn't be built on. Out there, there are no sounds but those of the wind and the sea and the birds.

It is curious how the sound of the wind moving over the water makes one sound, and the water sliding over the sand, and tumbling down over its own wave, forms another. The bird voice of these flats is the call of the willets. I have a new idea of these birds after seeing them here. I had always associated them with quiet water and salt marshes instead of the ocean beach. When I went down tonight one was standing at the edge of the water, looking out over it and giving its loud urgent cry. Presently there was an answer, and this bird flew to join the other — they greeted each other noisily and one flew off. [. . .]

These flats become even more wonderful as dusk approaches and the only light is that reflected from the occasional pools of water. Then bird forms become dark silhouettes, with no color discernable. Sanderlings scoot across the sand like little ghosts, and here and there, larger, darker forms of willets stand out. Often I could come very close to them before they would take alarm — the sanderlings by running, the willets by flying up, crying. Three black skimmers flew along the ocean edge while it was still light enough to see color.

As I was walking back in the near dark, I could see them flitting around like big moths. One "skimmed" along within a few yards of me, following a "creek" of water that wound across the flats. There seemed to be little fish in this — there were disturbances at the surface, sending little circular ripples spreading out.

Dunes

What peculiar brand of magic is inherent in that combination of sand and sky and water it is hard to say. It is bleak and stark. But somehow it is not forbidding. Its bleakness is part of its quiet, calm strength.

The dune land is a place of overwhelming silence, or so it seems at first. But soon you realize that what you take for silence is an absence of human created sound. For the dunes have a voice of their own, which you may hear if you will but sit down and listen to it. It is compounded of many natural sounds which are never heard in the roar of a city or even in the stir of a small town. A soft, confused, hollow rustling fills the air. In part it is the sound of surf on the beach half a mile away — a wide sand valley and another ridge of sand hills between us — the deep thunder of the surf reduced to a sigh by the intervening distance. In part it is the confused whisperings of the wind, which seems to be never wholly still there, but always to be exploring the contours of the land it made, roaming down into the valleys and leaping in little, unexpected gusts over the crests of the sand hills. And there are the smaller voices, the voices of the sand and the dune grass. The soft swish of the grass as it dips and bends in the wind to trace its idle arcs and circles in the sand at its feet. Arcs, especially on the southeast side of the grass, mean unsettled weather, so they say; whole circles foretell fair weather because they show the wind to be blowing alternately from different quarters. I cannot say as to that; but the scribblings of the dune grass always enchant me, though I cannot read their meaning.

I knew the history of that land, and there,

under the wind and beside the surf that had carved it, I recalled the story —

I stood where a new land was being built out of the sea, and I came away deeply moved. Although our intelligence forbids the idea, I believe our deeply rooted attitude toward the creation of the earth and the evolution of living things is a feeling that it all took place in a time infinitely remote. Now I understood. Here, as if for the benefit of my puny human understanding, the processes of creation — of earth building — had been speeded up so that I could trace the change within the life of my own contemporaries. The changes that were going on before my eyes were part and parcel of the same processes that brought the first dry land emerging out of the ancient and primitive ocean; or that led the first living creatures step by step out of the sea into the perilous new world of earth.

Water and wind and sand were the builders, and only the gulls and I were there to witness this act of creation.

Strange thoughts come to a man or woman who stands alone in that bleak and barren world. It is a world stripped of the gracious softness of the trees, the concealing mercies of abundant vegetation, the refreshment of a quiet lake, the beguilement of shade. It is a world stripped to the naked elements of life. And it is, after all, so newly born of the sea that it could hardly be otherwise. And then there is the voice of the sand itself — the quick sharp sibilance of a gust of sand blown over a dune crest by a sudden shift of the breeze, the all but silent sound of the never ending, restless shifting of the individual grains, one over another.

I am not sure that I can recommend the dunes as a tonic for all souls, nor for all moods. But I can say that anyone who will go alone into the dune lands for a day, or even for an hour, will never forget what he has seen and felt there.

∿ 18 ∿

[1953]

The Edge of the Sea

This paper, given the same title as her forthcoming book, was presented at the American Association for the Advancement of Science Symposium, "The Sea Frontier," and was the only purely scientific paper Carson ever gave to a professional academic organization. In it she pursues such broad ecological questions as "Why does an animal live where it does?" and "What is the nature of the ties that bind it to its world?" The paper reflects Carson's meticulous field research and the imagination with which she could apply the latest theoretical research on climate and temperature change to her general investigation of the evolution of life along the shoreline.

Although she defined herself as a scientist who wrote for the public, Carson could hold her own with the most advanced research biologists and

206

earn their respect. In a letter to a friend, however, she admitted she was uncharacteristically nervous about her presentation at this symposium because her mentor, Henry Bryant Bigelow, Harvard scientist and former director of the Woods Hole Oceanographic Laboratory, was in the audience. Carson later dedicated the 1961 edition of The Sea Around Us *to Bigelow.*

In recent years I have been dealing with the ecology of the seashore: with the animal and plant communities of the rocky coasts, the beach sands, the marshes and mud flats, the coral reefs and mangrove swamps. I have been thinking about the relations of one animal to other animals, of animals to plants, and of the animal or plant to the physical world about it. Always in such reflections one is made aware of the complex pattern of life. No thread is found to be complete in itself, nor does it have meaning alone. Each is but a small part of the intricately woven design of the whole, for the living organism is bound to its world by many ties, some of them relating to biology, others to chemistry, geology, or physics.

For example: perhaps we discover that an animal we have found year after year in the same place suddenly is found there no more, and eventually it appears that the whole population of this creature has shifted its range; that it has done this, moreover, in response to a change in a single factor of its physical environment — the temperature of the water.

Or we may find that a certain marine worm is so specialized in its living requirements that it can live in one kind of sand and one only; and moreover, that its very young stages — the larvae whose age is measured in hours or days — are able to find and to recognize sand of these particular qualities with a precision that few human students of geology could match.

Or again perhaps we discover a sudden and quite mysterious change in the ability of a marine animal to grow or to reproduce, or in the ability of its larvae to exist in the accustomed places, and perhaps we are led to suspect a subtle change in the biochemical nature of the seawater.

All or any of these things are enough to convince us that we, as biologists, cannot exist in a comfortable ivory tower of our own; that it is quite necessary that we concern ourselves with the related sciences if we are to understand the creatures of the marine world.

The edge of the sea is a laboratory in which Nature itself is conducting experiments in the evolution of life and in the delicate balancing of the living creature within a complex system of forces, living and non-living. We have come a long way from the early days of the biology of the shore, when it was enough to find, to describe, and to name the plants and animals found there. We have progressed, also, beyond the next period, the dawn age of the science of ecology, when it was realized that certain kinds of animals are typical of certain kinds of habitats. Now our minds are occupied with tantalizing questions. "Why does an animal live where it does?" "What is the nature of the ties that bind it to its world?"

One of the physical ties is especially interesting in this, our present period of earth history. It is a force that is omnipresent; no living thing is exempt from its influence. For life in the aggregate is lived within a relatively narrow range of temperature. The fact that our planet Earth has a fairly stable temperature helps to make it hospitable to life. In the sea, especially, temperature changes are gradual and moderate and many animals are so delicately adjusted that they cannot tolerate an abrupt or extensive change in the temperature of the surrounding water. If such occurs they must migrate or die.

Now our climate is changing and we are moving into a warm cycle of unknown duration. Ocean temperatures are slower to reflect the change than the air, but there has been a measurable warming of Atlantic coastal waters. The winter temperatures, which may be the critical ones for some marine animals, are less severe. Also, the water is warmer in summer. Records of winter temperatures for the Gulf of Maine (based on an average of records for Boston, Boothbay Harbor, Eastport, and St. Andrews) show an impressive rise. In 1918, for example, surface temperatures during the coldest month of the year averaged about 28°, and in the 1910–1919 decade, temperatures below 32° occurred almost every winter. During the 1930's, however, in only three years of the decade did the water temperatures of the coldest month average below 32°, and this has not happened in any year since 1940. Last winter's coldest month averaged 38°, or ten degrees warmer than the winter of 1918 — a very substantial increase as sea temperatures go.

It is not surprising that such a change in the oceanic climate should have caused many shifts of distribution among marine faunas. It is well known that the waters around Greenland and Iceland have been invaded by animals from regions to the south. This has been well documented in papers by Jensen and others. Less has been said about corresponding changes in the waters off the east coast of the United States. But many are occurring, and collectively they seem to form a significant picture.

Take, for example, the green crab. This is a member of the family of swimming crabs to which the blue crab also belongs. Once its range was very restricted. S. I. Smith, in Verrill's report on the Invertebrate Animals of Vineyard Sound, gave its distribution as "Cape Cod to New Jersey." More than a quarter of a century later, in 1905, Dr. Mary Rathbun placed its northward limit in Casco Bay. As late as 1930 she reported only one record of its occurrence east of Casco Bay. The 1929 biological survey of Mt. Desert Island did not include it. However, in 1930, two lots of green crabs were collected in the vicinity of Brooklin, in Hancock County, Maine. These were sent to the National Museum because they had been taken so far beyond the limits of their normal range. Nine years later, Leslie W. Scattergood [an FWS scientist of lobster culture] found the species at Winter Harbor, and in 1951 he was able to report its eastward spread to Lubec. A few months later he found it on the Maine shores of Passamaquoddy Bay, and Canadian biologists reported it from Oven Head on the eastern shores of this bay the same year.

According to the Biological Station at St. Andrews, the green crabs were very abundant on all the flats of Passamaquoddy Bay in the summer of 1953, when they were also found across on the Nova Scotian shores — in St. Mary Bay in August and in Minas Basin in November. There the record stands for the moment. The spread of the green crab is better documented than that of most species for a practical reason. Its invasion of the soft clam areas of Maine has, in some localities, almost wiped out the industry, for the crab preys so extensively on the young stages that farming of clams can hardly be practiced in its presence.

However, there are other occurrences perhaps equally interesting. For these largely unpublished records I am indebted to a number of biologists, including various members of the Fish and Wildlife Service, John [N. J.] Berrill of McGill University, and Fenner Chace of the National Museum.

With warming water temperatures, the sea herring is becoming scarce in Maine. Whether this is entirely a matter of temperature or whether diseases or other factors enter the picture I suppose no one can say with assurance. But as the herring declines other fish, members of the same family but of more southern distribution, are moving in. Back in the 1880's there was a substantial fishery for the menhaden at East Boothbay and some other Maine ports. Then the fish disappeared from Maine and for many years has been yielding enormous catches in New Jersey, Virginia, North Carolina, and other southern states. But about 1950 the menhaden came back into Maine waters in numbers, followed by Virginia boats and fishermen. Then there is another member of the same family, the round herring. In the 1920's Bigelow [Henry Bigelow of Harvard University] and Welsh reported it as occurring from the Gulf of Mexico north only as far as Cape Cod. But it was rare anywhere on the Cape, and two that had been caught at Provincetown were preserved in the Museum of Comparative Zoology. Now, however, immense schools of this fish have been appearing in Maine waters for several years, and the fishing industry is experimenting with canning it.

There are negative responses, too. One of the hydroids, a member of the genus *Syncoryne*, is so delicately adjusted to water temperatures that it has been considered by some a key temperature-zone organism. Professor [N. J.] Berrill tells me that 6 years ago this species was common at Ocean Point, near Boothbay Harbor, in June. So was the hydroid *Clava*. But two years later, in 1950,

Professor Berrill could find only a trace of *Syncoryne*, and since then none whatever, even in mid-winter. Apparently temperatures on this part of the coast have become too warm for it. The same appears to be true of *Clava*, of which none has been found during the past three years in the Boothbay region.

If we knew the whole story of each of these examples of a change of distribution we should very probably find the focus of our attention shifting to the larval stages of the animals that are involved. Often the adults might perfectly well be able to establish themselves in new areas outside of the normal temperature range of the species, but they cannot do so — that is, they cannot reproduce and have their young survive — because the waters are too cold or too warm for the welfare of the larvae. More and more it is becoming clear that the ecology of the adult marine animal is dependent upon the ecology of the larvae.

This fact is reinforced with beautiful clarity in another field of research, that has to do with the relations of the larval forms of some invertebrates to the substratum. This has been the subject of brilliant and significant work by Douglas Wilson at the Plymouth Laboratory.

Many invertebrates, as adults, live either

permanently or semi-permanently attached to the sea bottom, where rocks are exposed, or burrow into its covering of sand or mud. If such sedentary animals are to establish new colonies, this must be done by the larvae, for they alone have freedom to swim or even to be carried passively in the currents. The minute and delicate larvae often have a further responsibility, for many species are so specialized that they can inhabit only a certain kind of sea bottom. Sand, for example, is far from being a substance of uniform nature. It is diverse in its geologic origin, its chemical nature, its capacity to support life. One of the small annelid worms from which Douglas Wilson has learned so much about the reactions of larvae lives only in clean, coarse sand, stirred by strong tidal action, and composed mainly of quartz with some intermixture of materials derived from rocks and shells. Such sand occurs only in scattered areas on the shores of the English Channel, and the occurrence of this species of worm is accordingly limited.

To understand, then, how the adults come to live where they do, we must return to a study of those very early stages in the life of each member of the species, when clouds of young — the potential founders of new colonies — are launched into the sea with each spawning of the adults. Most of these larvae spend the early days — or, it may be, weeks — in the drifting community of the plankton, in the midst of diatoms, dinoflagellates, and other microscopic plants; in the company of minute crustaceans, worms, pteropods, and other permanent members of the plankton; and of hosts of other larvae, that, like themselves, are only temporary drifters and swimmers in the upper layers of the sea. Some of the larvae feed on the plant plankton, some on other larvae. Many are eaten by other members of the animal plankton, or are destroyed by cold or storms. Almost all are delicate, transparent, and minute, fragile as blown glass. Produced in astronomical numbers, they are destroyed with almost equal prodigality; seemingly the larvae are a tenuous link on which to base the security of the chain of existence.

But the larvae are not entirely without resources of their own, as we are discovering now from the work of Wilson and a few others. It seems that they have a fair amount of control over their own destinies, especially in that critical moment of life when they assume the form of the adult. Our early conception of this metamorphosis of the larva has been shown to be false for so many forms that there is some reason to believe we have been generally in error. We used to believe that this drastic change of form, from the larval to the adult stage, occurred at a certain moment in the life of the larva; and that it occurred at this moment whether or not the larva was at that time in surroundings suitable for taking up the adult existence. From these beliefs it would follow that a very large percentage of larvae would be lost because of being on unfavorable ground when the moment of metamorphosis arrived. However, thanks largely to the work of Wilson, we now have a new conception of this crisis in the life of the larva. In many forms, at least, we know that the larva has the ability to recognize the sands or muds of the type inhabited by its parents, that it may test out one area after another and may postpone its own metamorphosis for a considerable period of time, changing to the adult form only when a suitable substratum is found. A few sentences from one of Wilson's reports, on the tube-building polychaete, *Owenia fusiformis*, make this clear:

When a month old it can change suddenly, in a few seconds, from an object of elegance and beauty into an ugly little worm [. . .] busily engaged in swallowing the remains of its [. . .] larval organs. But, and this is the point, it will rarely do so successfully unless it be provided with sand of a suitable sort. [. . .] There is a period of about a week at any time during which it will metamorphose in contact with the sand in which the adult lives. The quick reaction to contact with such sand is strongly reminiscent of a chemical experiment; to a clean dish containing swimming larvae, sand is added, and almost at once there is a precipitate of worms.

From Wilson's experiments on this and other forms, we may visualize what happens when such a larva is ready for the choice of its adult home. Through its changing reactions to light it has perhaps already turned away from the surface waters and lives within the currents of water that flow over the sea bottom. Now and then it may drop down to the underlying bottom and enter the sand. But if the sand is found to be unsuitable — if it lacks the sought-for qualities — the larva emerges from it and enters again the slow drift of the currents, allowing itself to be carried on to new areas where, perchance, it will find that which it requires. When it does, the response is immediate; the larva settles down, and metamorphosis proceeds.

In learning this much, science has taken a

long step forward. There are still many questions that recur. What is the link between the delicate larva and its specialized physical environment? To what quality in the environment does it respond? What is the external stimulus that sets in motion those processes within the larva, transforming and remolding its tissues into the likeness of the adult?

One by one, these questions are being posed in the form [of] imaginatively contrived experiments. Wilson at first tested the possibility that the larvae react to sand grains of a certain size and shape. He concluded that although the grade of sand has some influence, it is not decisive. Then he considered whether a substance, possibly of organic nature, might be given off into the water by the native-type sands, attracting the larvae toward it. But it soon became clear that the larvae do not react, either negatively or positively, until they come into contact with the sand. In his most recently reported work, Wilson favors the theory that some organic material present on the surface of the sand grains causes them to attract or to repel the larvae. Further work along these lines is being done. In the meantime, it is established beyond question that some — perhaps most — species of marine bottom invertebrates have an inherited ability to recognize their own habitat when, as larvae, they first come into contact with it.

This fascinating subject is related to another that is in the forefront of biological thinking today. This is the subject of the so-called "ectocrines" — the products of metabolism that are liberated into the sea water by marine organisms. As yet no conceptions and no conclusions in this field are final; the subject lies on the misty borderlands of advancing knowledge. And yet almost everything that in the past we have taken for granted, or labeled as [an] insoluble problem, bears renewed thinking about in the light of what we know, or what we think is probable, about these substances of far-reaching effect.

In the sea there are mysterious comings and goings, both in space and time: the movements of migratory species, the strange phenomenon of succession by which, in one and the same area, one species appears in profusion, flourishes for a time, and then dies out, only to have its place taken by another and then another, like actors in a pageant passing before our eyes. And there are other mysteries. The phenomenon of "red tides" has been known from early days, recurring again and again down to the present time — a phenomenon in which the sea becomes discolored because of the extraordinary multiplication of some minute form, often a dinoflagellate, and in which there are disastrous side effects in the shape of mass mortalities among fish and some of the invertebrates. Then there is the problem of curious and seemingly erratic movements of fish, into or away from certain areas, often with sharp economic consequences. When the so-called "Atlantic water" floods the south coast of England, herring become abundant within the range of the Plymouth fisheries, certain characteristic plankton animals occur in profusion, and certain species of invertebrates flourish in the inter-tidal zone. When, however, this water mass is replaced by Channel water, the cast of characters undergoes many changes.

In these and other phenomena, the question recurs, and, unanswered, recurs again: Why? Here and there we perceive the first faint glimmerings of what may be the truth.

It appears that some, at least, of these things may in some measure be explained as the effects of substances present in the sea water — substances produced by one kind of organism as a by-product of its own metabolism, but exerting a powerful influence on another. A somewhat analogous and better known effect is that of antibiotic substances on bacteria. Apparently the ectocrines of the sea may be either harmful or beneficial in their effects. Of this much, however, science now seems certain: The nature of any mass of water, and its possible influence for good or ill upon the creatures it surrounds, are strongly influenced by the metabolism of those forms that, at an earlier point in time, lived within this same water.

It is interesting to trace the growth of an idea — a voice here, another there, and finally someone begins to put it all together, invents a terminology, and a new field of research is recognized. Probably it is only within the past decade that there has been much talk among biologists about external metabolites or ectocrines; yet this, also, seems to be one of the new-old subjects, for we can identify its embryonic beginnings in the literature of at least 70 years ago. In 1885 Pearcey reported in a Scottish journal his observation that herring are scarce in waters inhabited by certain diatoms, and that animal plankton also are scarce in such waters. About a quarter of a century ago, Johnstone, Scott, and Chadwick expressed their opinion that plankton communities influence each other, and that "there are what we may call group symbioses on the great scale, so that the kind of plankton which we may expect to be present in a certain sea area must depend, to some extent, on the kind of plankton that was previously present." Early in the thirties, Allee [W. C. Allee, animal ecologist at Woods Hole] made the significant statement that "aggregations of aquatic organisms condition the medium surrounding them by the addition of secretions and excreations, the nature of which forms one of the important problems of mass physiology."

While the existence of such organic substances in the sea now is generally recognized, there is little direct evidence as to their chemical nature or their precise role in the life processes of marine animals. Certain lines of investigation lead us back to the inshore waters and to the attached marine algae that grow in the coastal zone. These may be a source of very important ectocrines. If the possibilities now dimly foreseen are confirmed, these substances produced in coastal waters may act as catalytic agents to set off whole cycles of life in the sea. These waters are the habitat of the brown meadows of rockweeds, the dusky forests of the kelps, and the more fragile algae of pale green hue and delicate texture. These attached plants can live only as deep as light can penetrate and so are excluded from most of the open ocean.

In recent reports from the Goteborg Laboratory in Sweden we find that where the rockweeds *Fucus* and *Ascophyllum* are growing, the water acquires a property that stimulates the growth of the sea lettuce, *Ulva*, and also of *Enteromorpha*. From other work we know that the sea lettuce itself produces a substance that apparently is needed for the growth of certain diatoms in artificial media.

This is a plant-to-plant relation, but the ectocrines of the algae seem also to be concerned in an animal-plant relationship. In Japan, Miyazaki found that he could stimulate the spawning of oysters with a substance extracted from sea lettuce. This leads to a fascinating field of speculation. If indeed it is confirmed that ectocrines released into the sea by coastal vegetation induce both the flowering of the diatoms and the spawning of certain marine animals, a very neatly fitting chain of circumstances would result. The larval stages of many invertebrates, including oysters, feed on diatoms. The eggs of most lamelli-branchs develop into free-swimming plantonic larvae within a few days, so that one and the same stimulus could produce the young animals and the plants that will serve as their food.

A link between plant metabolites and animal reproduction is suggested by other observations. Rapidly maturing herring concentrate around the edges of patches of plant plankton, although the fully adult herring may avoid them. It has been suggested that "water-borne metabolites" influence the change of sex that regularly occurs in the mollusk *Crepidula*. The spawning adults, eggs, and young of some animals have been reported by Wimpenny [R. S. Wimpenny, a plankton expert] to occur more often in dense phytoplankton than in sparse patches. Others associate spawning of the copepod *Calanus* with dense phytoplankton. Recent research in the physiology of plant pigments seems significant in this connection, suggesting that the carotenoid pigments have a definite effect on sex and reproduction of animals.

So, even in the waters of the sea, we are brought back to the fundamental truth that nothing lives to itself. The water is altered, in its chemical nature and in its capacity for inducing metabolic change, by the fact that certain organisms have lived within it and by so doing have transmitted to it new properties with powerful and far-reaching effects. This is a field for imaginative and creative studies of the highest order, for in it we are brought face to face with one of the great mysteries of the sea.

∿ 19 ∿

[1954]

The Real World Around Us

The Sorority of Women Journalists, Theta Sigma Phi, invited Carson to speak about her experiences as a woman writer at its annual dinner in Columbus, Ohio, in the spring of 1954. With an audience of nearly a thousand women, Carson barely touched on the subject of her new book, The Edge of the Sea. *Instead she spoke more autobiographically than she had ever dared before.*

In the first part of her talk, Carson reflects on how she came to write about the sea, and her experiences sailing on it as a member of the crew of a U.S. Fish and Wildlife research vessel. The heart of her remarks, however, are devoted to her ideas about the meaning of life, particularly the crucial role natural beauty plays in the spiritual development of an individual or a society.

The audience, moved by the depth of Carson's

226

concern and obvious passion, gave her an enthu-siastic ovation, many women reaching out to press her hand as she left the hall. Although Carson never gave another speech of quite the same warmth and candor, its reception encouraged her to adopt a more personal style.

[. . .] I can remember no time, even in earliest childhood, when I didn't assume I was going to be a writer. I have no idea why. There were no writers in the family. I read a great deal almost from infancy, and I suppose I must have realized someone wrote the books, and thought it would be fun to make up stories, too.

Also, I can remember no time when I wasn't interested in the out-of-doors and the whole world of nature. Those interests, I know, I inherited from my mother and have always shared with her. I was rather a solitary child and spent a great deal of time in woods and beside streams, learning the birds and the insects and flowers.

There is another thing about my childhood that is interesting now, in the light of later happenings. I might have said, with Emily Dickinson:

I never saw a moor,

I never saw the sea;
Yet know I how the heather looks,
And what a wave must be.

For I never saw the ocean until I went from college to the marine laboratories at Woods Hole, on Cape Cod. Yet as a child I was fascinated by the thought of it. I dreamed about it and wondered what it would look like. I loved Swinburne and Masefield and all the other great sea poets.

I had my first prolonged contact with the sea at Woods Hole. I never tired of watching the tidal currents pouring through the Hole — that wonderful place of whirlpools and eddies and swiftly racing water. I loved to watch the waves breaking at Nobska Point after a storm. At Woods Hole, too, as a young biologist, I first discovered the rich scientific literature of the sea. But it is fair to say that my first impressions of the ocean were sensory and emotional, and that the intellectual response came leter.

Before that meeting with the sea had been accomplished, however, I had a great decision to make. At least, I thought I had. I told you that I had always planned to be a writer; when I went to college, I thought the way to accomplish that was to major in English composition. Up to that time, despite my love for the world of nature, I'd had no training in biology. As a college sophomore, I was exposed to a fine introductory course in biology, and my allegiance began to waver. Perhaps I wanted to be a scientist. A year later the decision for science was made; the writing courses were abandoned. I had given up writing forever, I thought. It never occurred to me that I was merely getting something to write about. What surprises me now is that apparently it didn't occur to any of my advisors, either.

The merging of the two careers didn't begin until several years after I had left Johns Hopkins, where I had gone to do graduate work in zoology. Those were depression and post-depression years, and after a period of part-time teaching jobs, I supplemented them with another part-time assignment. The Bureau of Fisheries in Washington had undertaken to do a series of radio broadcasts. They were looking for someone to take over writing the scripts — someone who knew marine biology and who also could write. I happened in one morning when the chief of the biology division was feeling rather desperate — I think at that point he was having to write the scripts himself. He talked to me a few minutes and then said: "I've never seen a written word of yours, but I'm going to take a sporting chance."

That little job, which eventually led to a permanent appointment as a biologist, was in its way a turning point. One week I was told to produce something of a general sort about the sea. I set to work, but somehow the material rather took charge of the situation and turned into something that was, perhaps, unusual as a broadcast for the Commissioner of Fisheries. My chief read it and handed it back with a twinkle in his eye. "I don't think it will do," he said. "Better try again. But send this one to the *Atlantic*." Eventually I did, and the *Atlantic* accepted it. Since then I have told my chief of those days that he was really my first literary agent.

From those four *Atlantic* pages, titled "Undersea," everything else followed. Quincy Howe, then editor for Simon and Schuster, wrote to ask why I didn't do a book. So did Hendrik Willem van Loon. My mail had never contained anything so exciting as his first letter. It arrived in an envelope splashed with the green waves of a sea through which van Loon sharks and whales were poking inquiring snouts.

That was only the beginning of a wonderful correspondence, for it seemed Hendrik van Loon had always wanted to know what lay undersea, and he was determined I should tell the world in a book or books. His typing was amazing but his handwritten letters were almost illegible. Often he substituted a picture for a word, and that helped. After a few weeks of such correspondence, I spent a few days with the van Loons in their Connecticut home, during which I was properly introduced to my future publisher.

To a young and very tentative writer, it was a stimulating and wonderful thing to have the interest of this great man, so overwhelming in his person and his personality, but whose heart was pure gold. Through him, I had glimpses of a world that seemed exciting and fabulous, and I am sure his encouragement had a great deal to do with the fact that my first book, *Under the Sea-Wind*, was eventually published.

When that happened, however, on the eve of Pearl Harbor, the world received the event with superb indifference. The reviewers were kind, but that rush to the book store that is the author's dream never materialized. There was a Braille edition, a German translation, and use of various chapters in anthologies. That was all. I was busy with war work, and when I thought at all about writing, it was in terms of magazine pieces; I doubted that I would ever write another book. But I did, and ten years after *Under the Sea-Wind, The Sea Around Us* was published.

The fifteen years that I spent in fishery and wildlife conservation work with the Government have taken me into certain places where few other women have been. Perhaps you would like to hear about some of those.

While I was doing information work for Fish and Wildlife, the Service acquired a research vessel for work at sea, specifically on the famous fishing ground known as Georges Bank, that lies some 200 miles east of Boston and south of Nova Scotia. Some of the valuable commercial fishes are becoming scarce on the Bank, and the Service is trying to find the reason. *The Albatross III*, as this converted fishing trawler was called, operated out of Woods Hole, making repeated trips to Georges. She was making a census of the fish population; this was done by fishing according to a systematic plan over a selected series of stations. Of course, various scientific data on water temperatures and other matters were collected, too.

It was decided finally — and I might have had something to do with originating the idea — that perhaps I could do a better job of handling publications about the *Albatross* if I had been out on her. But there was one great obstacle. No woman had ever been on the *Albatross*. Tradition is important in the Government, but fortunately I had conspirators who were willing to help me shatter precedent. But among my male colleagues who had to sign the papers, the thought of one woman on a ship with some fifty men was unthinkable. After much soul searching, it was decided that maybe *two* women would be all right, so I arranged with a friend, who was also a writer, to go with me. Marie [Rodell] thought she would write a piece about her experiences, and declared that her title would be: "I Was a Chaperone on a Fishing Boat."

And so one July day we sailed from Woods Hole into ten days of unusual adventure. This is not the place to tell about the scientific work that was done — but there was a lighter side, especially for us who were mere observers, and there were unforgettable impressions of fishing scenes; of fog on Georges, where the cold water and the warm air from the Gulf Stream are perpetually at war at that season of the year; and of the unutterable loneliness of the sea at night as seen from a small vessel.

As to the lighter side — a fishing trawler is not exactly a luxury liner, and both of us were on our mettle to prove that a woman could take it without complaining. Hardly had the coast of Massachusetts disappeared astern when some of the ship's officers began to give us a vivid picture of life aboard. The *Albatross*,

233

they told us, was a very long and narrow ship and rolled like a canoe in a sea, so that everyone got violently seasick. They described some of the unpleasant accidents that sometimes occur in handling the heavy gear. They told us about the bad food. They made sure we understood that the fishing process went on night and day, and that it was very noisy.

Well — not all the things those Job's comforters predicted came true, but a great many of them did. However, we learned in those ten days that one gets used to almost anything.

We learned about the fishing the very first night. After steaming out through Nantucket Channel late in the afternoon, we were to reach our first fishing station about midnight. Marie and I had gone to bed and were sound asleep when we heard a crash, presumably against the very wall of our cabin, that brought us both upright in our bunks. Surely we had been rammed by another vessel. Then a series of the most appalling bangs, clunks, and rumbles began directly over our heads, a rhythmic thundering of machinery that would put any boiler factory to shame. Finally it dawned upon us that this was fishing! It also dawned on us that this was what we had to endure for the next ten nights. If there had been any way to get off the *Albatross* then I'm sure we would have taken it.

At breakfast the next morning there were grins on the faces of the men. "Hear anything last night?" they asked. Both of us wore our most demure expressions. "Well," said Marie, "once we thought we heard a mouse, but we were too sleepy to bother." They never asked us again. And after a night or two we really did sleep through the night.

One of the most vivid impressions I carried away from the *Albatross* was the sight of the net coming up with its load of fish. The big fishing trawlers such as this one drag a cone-shaped net on the floor of the ocean, scraping up anything lying on the bottom or swimming just above it. This means not only fish but also crabs, sponges, starfish and other life of the sea floor. Much of the fishing was done in depths of about 100 fathoms, or 600 feet. After a half hour of trawling the big winches would begin to haul in the cables, winding them on steel drums as they came aboard. There is a marker on every hundred fathoms of cable, so one can tell when to expect the big net to come into view, still far down in the green depths.

I think that first glimpse of the net, a shapeless form, ghostly white, gave me a sense of sea depths that I never had before. As the net rises, coming into sharper focus, there is a stir of excitement even among the experienced fishermen. What has it brought up?

No two hauls are quite alike. The most interesting ones came from the deeper slopes. Georges Bank is like a small mountain resting on the floor of a surrounding deeper sea — most of the fishing is done on its flat plateaus, but sometimes the net is dragged down on the slopes near the mountain's base. Then it brings up larger fish from these depths. There is a strange effect, caused by the sudden change of pressure. Some of the fish become enormously distended and float helplessly on their backs. They drift out of the net as it nears the surface but they are quite unable to swim down.

Then one sees the slender shapes of sharks moving in to the kill. There was something very beautiful about those sharks to me — and when some of the men got out rifles and killed them for "sport" it really hurt me.

In those deep net hauls, too, there were often the large and grotesque goosefish or angler fish. The angler has a triangular shape, and its enormous mouth occupies most of the base of the triangle. It lives on the floor of the sea, preying on other fish. The anglers always seemed to have been doing a little fishing of their own as the net came up, and sometimes the tails of two or three large cod would be protruding from their mouths.

Sometimes at night we would go up on the deck to watch the fishing. Then the white splash of electric light on the lower deck was the only illumination in a world of darkness and water. It was a colorful sight, with the men in their yellow oilskins and their bright flannel shirts, all intensified and made somehow dramatic by the blackness that surrounded them.

There is something deeply impressive about the night sea as one experiences it from a small vessel far from land. When I stood on the afterdeck on those dark nights, on a tiny man-made island of wood and steel, dimly seeing the great shapes of waves that rolled about us, I think I was conscious as never before that ours is a water world, dominated by the immensity of the sea.

However, it is a curious thing that one sometimes experiences a sense of the sea on land. A few years ago I had a wonderful opportunity to go far into the interior of the Everglades in Florida. Many people have crossed this great wilderness by way of the Tamiami Trail. That is better than not seeing it at all, but until one has penetrated far into the interior, into the trackless, roadless areas of the great swamp, one does not know the Everglades.

The difficulties of travel there are great, and no ordinary means of transportation will do. But a few pioneering individuals have developed wonderful vehicles called "glades buggies." They were first used, I believe, to prospect for oil in the interior of the Everglades. They are completely independent of roads; they can go through water, they can navigate the seas of "saw-grass" or even push through low-growing thickets of trees and shrubs; they can make their way — painfully but surely — over ground pitted with holes and strewn with jagged boulders.

I learned about the glades buggies when I was on a trip for my office to the area that is now the Everglades National Park. At that time the Fish and Wildlife Service had responsibility for protecting the wildlife of the area. Two of us were staying at a hotel in Miami Beach, visiting various wildlife areas in the vicinity. When we heard about Mr. Don Poppenhager and his wonderful glades buggy, we decided to try to arrange a trip.

Mr. Poppenhager had never taken a woman into the swamp and at first he was hesitant. He warned us that it was a very uncomfortable experience; we assured him we could take it and really wanted to go. So he agreed to meet us at a little store on the

Tamiami Trail kept by a character known as Ma Szady.

I think our elegant Miami Beach hotel had been a little suspicious of our comings and goings on strange errands and in strange costumes, but the morning we left for the Everglades trip was almost too much for them. One of the Fish and Wildlife men was to pick us up at 5 A.M. and take us over the trail. This was in the summer, and a tropical darkness still hung over Miami at that hour. Not wanting to arouse the hotel, Shirley [Briggs] and I crept down the stairs laden with all our strange gear. As we tiptoed through the lobby, the head of a very sleepy but thoroughly suspicious clerk rose above the desk. "Are you ladies checking out?" he asked. I don't think his estimate of us rose when a very noisy, two-ton Government truck roared down the street and stopped at the hotel for its passengers.

The glades buggy that was waiting for us was a wonderful conveyance. It was built something like a tractor, with six pairs of very large wheels. Its engine was completely naked and exposed, and during the trip blasted its heat on the three of us perched on the buggy's single seat. There were various tools — pliers, screwdrivers, etc. — in a little rack against the motor block, and from time to time Mr.

Poppenhager leaned out as we jogged along and turned something or jabbed at the motor. It seemed to be in a perpetual state of boiling over; and now and then Mr. P. would stop and get out with a tin can and dip up some water — there was water everywhere — and pour it into the radiator. Usually he would drink a little — "the best water in the world" he would say.

But as I said a while ago, there was a curious sense of the sea there in the heart of the Everglades. At first I couldn't analyze it but I felt it strongly. There is first of all a sense of immense space from the utter flatness of the land and the great expanse of sky. The feeling of space is almost the same as at sea. The cloud effects were beautiful and always changing, and rain came over the grass, making a beautiful soft play of changing color — all grey and soft green. And again I found myself remembering rain at sea, dimpling the soft grey sheet of water. And in the Everglades the coral rock is always cropping out — underlying the water and raised in jagged boulders among the grass. Once that rock was formed by coral animals, living in a shallow sea that covered this very place. There is today the feeling that the land has formed only the thinnest veneer over this underlying platform of the ancient sea — that at any time

240

the relations of sea and land might again be reversed.

And as we traveled from one to another of the "hammocks" of palmetto and other trees that rise here and there in the great sea of grass, we thought irresistibly of islands in the ocean. Except for scattered cypresses, all the trees of this part of the Everglades are concentrated in the hammocks, which form where depressions in the rock accumulate a little soil. Everywhere else there is only rock, water, and grass. The hammocks are famous for their tree snails, which live on certain locust-like trees, feeding on mossy growths on the bark. The shells of the tree snails are brightly colored, with an amazing variety of patterns. They are so much sought by collectors that the more accessible hammocks have been stripped bare. On our steaming iron monster, we rode along through the hammocks, passing under the trees and picking off tree snails, as in childhood we used to snatch the iron rings on a merry-go-round.

During the day we went calling on several alligators known by Mr. Poppenhager to inhabit certain "holes." The first one was not at home; the second was. He was apparently out in his front yard, but at our approach he went crashing through the willows and into his pond. In the Everglades, a " 'gator hole" is typically a water-filled depression in the middle of a small hammock. Usually there is a rocky cave in the floor of this pond to which the alligator can retreat.

The Everglades is, of course, the land of the Seminole Indians. Far in the interior of the Glades we visited the sites of two ancient Indian villages. Some of these are being studied by archeologists who have found evidence of early tribes who antedated the Seminoles by several hundred years. Near one of the modern settlements, Mr. Poppenhager took us to visit an Indian grave. Because of the solid limestone floor of this whole region, there is no burial in the ordinary sense; the coffin is placed on the ground and the man is given his gun and other equipment he will need for his life in the next world.

To us the whole area seemed as trackless and as lacking in landmarks as the sea, but our guide knew exactly where he was going. Our only bad moments came late in the afternoon, when there began to be some question whether we had enough gas to get us back to the Trail. Mosquitoes had been with us all day, settling in clouds every time we stopped moving. So the thought of a night in the swamp wasn't pleasant. However, we made it about dusk, just as the Game Warden and the Fish and Wildlife patrolman were beginning to line up cars along the Trail to guide us back by their headlights.

That Fish and Wildlife patrolman was such an unforgettable character that I must tell you a little about him before we leave the Everglades. As Service patrolman for the area, it was his job to protect the birds and alligators and other wildlife from being molested. That meant he had to live far out in a wild part of the Everglades, where days went by without his seeing another person. The Service had had trouble filling the job. There were few men that would have taken it; and perhaps no one else as beautifully fitted for it as Mr. Finneran. He was tired of the northern cities where he had spent most of his life, and for about ten years he had known this wilderness of southern Florida. He had somehow gained the confidence of the Seminoles, who ordinarily have no love for the white man. But they admired and trusted Mr. Finneran — so much that they had given him a name and practically adopted him into their tribe. When the Service offered Mr. Finneran this lonely job, he took it gladly, and moved into the little shack that was to serve as home and headquarters. There he lived with a little dog, a few chickens, and a blue indigo snake named Chloe. He had five tree snails on a tree beside the house. He was very proud of them, and when we returned from our glades buggy trip, we brought him a few snails as a gift. He couldn't have been more pleased if they had been pure gold. I remember how feelingly he spoke to me of the beauty of the Everglades in the early morning, with dew on the grass and thousands of spider webs glistening. He spoke of the birds coming in such numbers they were like dark clouds in the sky. He told of the eerie silver light of the moon, and the red, glowing hordes of alligators in the ponds. His paradise had its flaws, as he acknowledged. He couldn't have a light in his shack at night because of the terrible Glades mosquito. Sometimes, on rainy nights, fire ants invaded the house and even swarmed into his bed. The Indians said ghosts haunted the place because it was built on an old Indian mound; but Mr. Finneran had heard no ghosts he couldn't explain. When city dwellers asked him how he stood the loneliness out there, he always asked how they endured sitting around in night clubs. "I wouldn't trade my life for anything," he told us.

From what I have told you, you will know that a large part of my life has been concerned with some of the beauties and mysteries of this earth about us, and with the even greater mysteries of the life that inhabits it. No one can dwell long among such subjects without thinking rather deep thoughts, without asking himself searching and often unanswerable questions, and without achieving a certain philosophy.

There is one quality that characterizes all of us who deal with the sciences of the earth and its life — we are never bored. We can't be. There is always something new to be investigated. Every mystery solved brings us to the threshold of a greater one.

I like to remember the wonderful old Swedish oceanographer, Otto Petterson. He died a few years ago at the age of 93, in full possession of his keen mental powers. His son, also a distinguished oceanographer, tells us in a recent book how intensely his father enjoyed every new experience, every new discovery concerning the world about him. "He was an incurable romantic," the son wrote, "intensely in love with life and with the mysteries of the Cosmos which, he was firmly convinced, he had been born to unravel." When, past 90, Otto Petterson realized he had not much longer to enjoy the earthly scene, he said to his son: "What will sustain me in my last moments is an infinite curiosity as to what is to follow."

The pleasures, the values of contact with the natural world, are not reserved for the scientists. They are available to anyone who will place himself under the influence of a lonely mountain top — or the sea — or the stillness of a forest; or who will stop to think about so small a thing as the mystery of a growing seed.

I am not afraid of being thought a sentimentalist when I stand here tonight and tell you that I believe natural beauty has a necessary place in the spiritual development of any individual or any society. I believe that whenever we destroy beauty, or whenever we substitute something man-made and artificial for a natural feature of the earth, we have retarded some part of man's spiritual growth.

I believe this affinity of the human spirit for the earth and its beauties is deeply and logically rooted. As human beings, we are part of the whole stream of life. We have been human beings for perhaps a million years. But life itself — passes on something of itself to other life — that mysterious entity that moves and is aware of itself and its surroundings, and so is distinguished from rocks or senseless clay — [from which] life arose many hundreds of millions of years ago. Since then it has developed, struggled, adapted itself to its surroundings, evolved an infinite number of forms. But its living protoplasm is built of the same elements as air, water, and rock. To these the mysterious spark of life was added.

Our origins are of the earth. And so there is in us a deeply seated response to the natural universe, which is part of our humanity.

Now why do I introduce such a subject tonight — a serious subject for a night when we are supposed to be having fun? First, because you have asked me to tell you something of myself — and I can't do that without telling you some of the things I believe in so intensely.

Also, I mention it because it is not often I have a chance to talk to a thousand women. I believe it is important for women to realize that the world of today threatens to destroy much of that beauty that has immense power to bring us a healing release from tension. Women have a greater intuitive understanding of such things. They want for their children not only physical health but mental and spiritual health as well. I bring these things to your attention tonight because I think your awareness of them will help, whether you are practicing journalists, or teachers, or librarians, or housewives and mothers.

What are these threats of which I speak? What is this destruction of beauty — this substitution of man-made ugliness — this trend toward a perilously artificial world? Unfortunately, that is a subject that could require a whole conference, extending over many days. So in the few minutes that I have to devote to it, I can only suggest the trend.

We see it in small ways in our own communities, and in larger ways in the community of the state of the nation. We see the destruction of beauty and the suppression of human individuality in hundreds of suburban real estate developments where the first act is to cut down all the trees and the next is to build an infinitude of little houses, each like its neighbor.

We see it in distressing form in the nation's capital, where I live. There in the heart of the city we have a small but beautiful woodland area — Rock Creek Park. It is a place where one can go, away from the noise of traffic and of man-made confusions, for a little interval of refreshing and restoring quiet — where one can hear the soft water sounds of a stream on its way to river and sea, where the wind flows through the trees, and a veery sings in the green twilight. Now they propose to run a six-lane arterial highway through the heart of that narrow woodland valley — destroying forever its true and immeasurable value to the city and the nation.

Those who place so great a value on a highway apparently do not think the thoughts of an editorial writer for the *New York Times* who said: "But a little lonesome space, where nature has her own way, where it is quiet enough at night to hear the patter of small paws on leaves and the murmuring of birds, can still be afforded. The gift of tranquillity, wherever found, is beyond price."

We see the destructive trend on a national scale in proposals to invade the national parks with commercial schemes such as the building of power dams. The parks were placed in trust for all the people, to preserve for them just such recreational and spiritual values as I have mentioned. Is it the right of this, our generation, in its selfish materialism, to destroy these things because we are blinded by the dollar sign? Beauty — and all the values that derive from beauty — are not measured and evaluated in terms of the dollar.

Years ago I discovered in the writings of the British naturalist Richard Jefferies a few lines that so impressed themselves upon my mind that I have never forgotten them. May I quote them for you now?

The exceeding beauty of the earth, in her splendor of life, yields a new thought with every petal. The hours when the mind is absorbed by beauty are the only hours when we really live. All else is illusion, or mere endurance.

Those lines are, in a way, a statement of the creed I have lived by, for, as perhaps you have seen tonight, a preoccupation with the wonder and beauty of the earth has strongly influenced the course of my life.

Since *The Sea Around Us* was published, I have had the privilege of receiving many letters from people who, like myself, have been steadied and reassured by contemplating the long history of the earth and sea, and the deeper meanings of the world of nature. These letters have come from all sorts of people. There have been hairdressers and fishermen and musicians; there have been classical scholars and scientists. So many of them have said, in one phrasing or another: "We have been troubled about the world, and had almost lost faith in man; it helps to think about the long history of the earth, and of how life came to be. And when we think in terms of millions of years, we are not so impatient that our own problems be solved tomorrow."

In contemplating "the exceeding beauty of the earth" these people have found calmness and courage. For there is symbolic as well as actual beauty in the migration of birds; in the ebb and flow of the tides; in the folded bud ready for the spring. There is something infinitely healing in these repeated refrains of nature — the assurance that dawn comes after night, and spring after winter.

Mankind has gone very far into an artificial world of his own creation. He has sought to insulate himself, with steel and concrete, from the realities of earth and water. Perhaps he is intoxicated with his own power, as he goes farther and farther into experiments for the destruction of himself and his world. For this unhappy trend there is no single remedy — no panacea. But I believe that the more clearly we can focus our attention on the wonders and realities of the universe about us, the less taste we shall have for destruction.

∿ **20** ∿

[1956]

Biological Sciences

Carson agreed to contribute an essay and select a bibliography on the biological sciences for Good Reading, *a reference book sponsored by the National Council of the Teachers of English. Although the Council paid contributors handsomely, Carson found that putting together the bibliography of the best books in the biological sciences required a great deal more work than she had anticipated.*

Her brief introduction to the books she finally selected provides insight into her attitudes toward science in general and to biology in particular at a time when she anticipated beginning her own research for a book on evolutionary biology.

Carson emphasizes the new science of ecology in her definition of the scope of the biological sciences, reinforcing her view that "nothing lives

252

unto itself." Lamenting the remoteness of science from the average citizen, Carson characteristically recommended that students explore their subjects first in nature and in the writings of the great naturalists before venturing into the laboratory.

The scope of biology can be truly defined only in broad terms as the history of the earth and all its life — the past, the present, and the future. Any definition of lesser scope becomes narrow and academic and fails utterly to convey the majestic sweep of the subject in time and space, embracing all that has made man what he is, and holding a foretaste of what he may yet become. For it has dawned upon us in these recent years of the maturing of our science that neither man nor any other living creature may be studied or comprehended apart from the world in which he lives; that such restricted studies as the classification of plants and animals or descriptions of their anatomy and physiology (upon which the early biologists necessarily focused their attention) are but one small facet of a subject so many-sided, so rich in beauty and fascination, and so filled with significance that no informed reader can neglect it.

In the truest sense, there is no separate literature of biology or of any science. Knowledge of the facts of science is not the prerogative of a small number of men, isolated in their laboratories, but belongs to all men, for the realities of science are the realities of life itself. We cannot understand the problems that concern us in this, our particular moment of time, unless we first understand our environment and the forces that have made us what we are, physically and mentally.

Biology deals with the living creatures of the living earth. Pleasure in color, form, and movement, awareness of the amazing diversity of life, and the enjoyment of natural beauty are part of man's heritage as a living creature. Our first conscious acquaintance with the subject should come, if possible, through nature — in fields and forests and on the shore; secondarily and by way of amplification and verification we should then explore its laboratory aspects. Some of the most gifted and imaginative biologists have first approached their subject through the medium of sensory impression and emotional response. The most memorable writings — though they be addressed to the intellect — are rooted in man's emotional reaction to that life stream of which he is a part. The writing of the great naturalists such as Hudson and Thoreau, most easily sampled in some of the excellent anthologies now available, has a valid place in one's reading in the field of biology.

As the frontiers of science expand, there is inevitably an increasing trend toward specialization, in which all the mental faculties of a man or group of men are brought to bear upon a single aspect of some problem. But there is fortunately a counter tendency, which brings different specialists together to work in cooperation. Oceanographic expeditions commonly include biologists, chemists, physicists, geologists, and meteorologists, so diverse are the problems presented by one aspect of the earth's surface. Atomic physicists, by discovering that radioactive elements in fossils and minerals disintegrate at a rate that may be determined, have provided biologists with a tool that has already revolutionized our concept of the age of the earth and permits a far more accurate approach than ever before to the problem of the evolution of man himself. Chemists and geneticists, by joining forces, seem to be solving the riddle of the gene and the actual means by which it produces hereditary characteristics.

Only within the 20th Century has biological thought been focused on ecology, or the relation of the living creature to its environment. Awareness of ecological relationships is

— or should be — the basis of modern conservation programs, for it is useless to attempt to preserve a living species unless the kind of land or water it requires is also preserved. So delicately interwoven are the relationships that when we disturb one thread of the community fabric we alter it all — perhaps almost imperceptibly, perhaps so drastically that destruction follows.

If we have been slow to develop the general concepts of ecology and conservation, we have been even more tardy in recognizing the facts of the ecology and conservation of man himself. We may hope that this will be the next major phase in the development of biology. Here and there awareness is growing that man, far from being the overlord of all creation, is himself part of nature, subject to the same cosmic forces that control all other life. Man's future welfare and probably even his survival depend upon his learning to live in harmony, rather than in combat, with these forces. [. . .]

❧ 21 ❧

[1956]

Two Letters to Dorothy
and Stanley Freeman

Carson shared most of her experiences exploring the tide pools and rocky shores near her cottage on Southport Island with her summer neighbors Dorothy and Stanley Freeman. With Dorothy in particular, Carson found a kindred spirit of deep emotional significance. In 1956 Carson's mother was an invalid, and Carson's niece Marjorie and her four-year-old son Roger, whom Carson later adopted after Marjorie's death, had come to Maine for a visit. Carson's account of their midnight exploration of the spring tide was written in a letter to the Freemans, who were not in Maine at the time.

Similarly, an October sunset produced the backdrop for a great migration of waterfowl across the horizon which moved Carson once again to share her experience with the absent Freemans.

257

Dear Stan and Dorothy

This morning I achieved the difficult feat of getting up without disturbing anyone but Jeffie, so maybe I can write a letter before breakfast.

Knowing you can't be at Southport as soon as you want to be, I'm always of two minds now about talking of the place or telling you anything special that happens — should I share it with you, or is it mean to talk about things you want so badly to see or do yourselves? That, in general, is my predicament, but this time I *have* to tell you about something strange and wonderful.

We are now having the spring tides of the new moon, you know, and they have traced their advance well over my beach the past several nights. Roger's raft has to be secured by a line to the old stump, so Marjie and I have an added excuse to go down at high tide. There had been lots of swell and surf and noise all day, so it was most exciting down there toward midnight — all my rocks crowned with foam, and long white crests running from my beach to Mahard's. To get the full wildness, we turned off our flashlights — and then the real excitement began. Of course, you can guess — the surf was full of diamonds and emeralds, and was throwing them on the wet sand by the dozen. Dorothy dear — it was the night we were there all over, but with everything intensified. A wilder accompaniment of noise and movement, and a great deal more phosphorescence. The individual sparks were so large — we'd see them glowing in the sand, or sometimes, caught in the in-and-out play of water, just riding back and forth. And several times I was able to scoop one up in my hand in shells and gravel, and think surely it was big enough to see — but no such luck.

Now here is where my story becomes different. Once, glancing up, I said to Marjie jokingly, "Look — one of them has taken to the air!" A firefly was going by, his lamp blinking. We thought nothing special of it, but in a few minutes one of us said, "There's that firefly again." The next time he really got a reaction from us, for he was flying so low over the water that his light cast a long surface reflection, like a little headlight. Then the truth dawned on me. He "thought" the flashes in the water were other fireflies, signaling to him in the age-old manner of fireflies! Sure enough, he was soon in trouble and we saw his light flashing urgently as he was rolled around in the wet sand — no question this time which was insect and which the unidentified little sea will-o-the-wisps!

You can guess the rest: I waded in and rescued him (the will-o-the-wisps had already had me in icy water to my knees so a new wetting didn't matter) and put him in Roger's bucket to dry his wings. When we came up we brought him as far as the porch — out of reach of temptation, we hoped.

It was one of those experiences that gives an odd and hard-to-describe feeling, with so many overtones beyond the facts themselves. I have never seen any account scientifically, of fireflies responding to other phosphorescence. I suppose I should write it up briefly for some journal if it actually isn't known. Imagine putting that in scientific language! And I've already thought of a child's story based on it — but maybe that will never get written.

Then everyone got up, and the day began! [. . .]

Dear Dorothy and Stan,

I hope this may reach you on your anniversary, but whenever it comes I know you will accept it as a little observance of that occasion. You know this is the first year since I have really known you that I have had to *write* in order to wish you happiness on the day, and many years of continuing happiness together. Having shared Your Day to some extent for the past two years, it has become a sort of Anniversary for me, too, with deep meanings that I know you understand without my putting them into words.

There are certain events that I've come to associate with the week — if not the actual day — of your anniversary, and now I must tell you what happened Friday evening. It had been one of those bright, clear days with a piercing wind from the northwest, and at sunset there was not a cloud in the sky. There had been a thought in my mind all day, and shortly after sunset I went into the living room and began to scan the horizon. Almost instantly I saw a faint line like a wisp of smoke above the Kennebec — then more and more until I knew that one of those great migrations of waterfowl was moving toward Merrymeeting Bay. All, as far as I saw, were far away in the western sky, but with the glasses their formations and even the individual birds stood out clearly. And the flights continued until dusk made the drifting ribbons invisible. One more detail: I had also had in mind that on that evening I should see the new moon — the moon of the month in which I must leave here for another season. But when I looked into that clear, after-sunset sky I couldn't see it. Behind the spruces on the far shore of the Bay the sky was a pale orange, fading above into yellow and then a cold, gray-blue. Then the ducks appeared, and as I was searching the sky with my glasses, suddenly I saw the moon just above the horizon, a thin sickle, but so enormous that at first I could hardly believe it actually was the moon! Its color was so close to that of the sky that without the glasses I couldn't see it. Last night it was clear in the evening sky, and soon, I suppose, I can begin to watch for its reflection in the Bay.

ᕦ 22 ᕤ

[1956]

The Lost Woods. A Letter to Curtis and Nellie Lee Bok

Deeply involved in organizing the Maine Chapter of the Nature Conservancy in the summer of 1956, Carson had preservation issues much on her mind. Through her friendship with Curtis Bok, President Judge of the Pennsylvania Supreme Court, whose family foundation had established the Mountain Lake Sanctuary in Mountain Lake, Florida, Carson had seen first-hand how effective personal philanthropy could be in saving beautiful places.

That fall Carson spent a windy morning exploring the shore and adjacent woods some distance north of her property. She and Dorothy Freeman called the area the "Lost Woods," after the title of a favorite essay by the English naturalist H. M. Tomlinson. She wrote to Dorothy,

263

If only [the land] could be kept always just as it is! If ever I wish for money — lots of it — it is when I see something like that. [. . .] Just for fun, tell me what you think, and let's pretend we could somehow create a sanctuary there, where people like us could go, as my friend said of the Bok Tower and grounds, "and walk about, and get what they need." Well, if no one ever thinks of it, it certainly won't happen; if someone does think hard enough, it just might.

Carson felt she now had a model, at least in spiritual intent, of how the Lost Woods might become a sanctuary, if she could just put enough money together from her future writing. Energized by this idea, Carson wrote to the Boks asking for advice on how to proceed.

Although the purchase price eventually proved beyond her means, Carson's dream has been fulfilled, and a large part of the shore she loved is now protected through the efforts of the Boothbay Regional Land Trust.

Dear Curtis and Nellie Lee,

[. . .] I think you understand this in me, even though we've had little chance to talk about it — my feeling for whatever beautiful and untouched oases of natural beauty remain in the world, my belief that such places can bring those who visit them the peace and spiritual refreshment that our "civilization" makes so difficult to achieve, and consequently my conviction that whenever and wherever possible, such places must be preserved. [. . .]

When a few years back and for the first time in my life, money somewhat beyond actual needs began to come to me through *The Sea Around Us*, I felt that, almost above all else, I wished some of the money might go, even in a modest way, to furthering these things I so deeply believe in. [. . .]

[The Lost Woods'] charm for me lies in its combination of rugged shore rising in rather steep cliffs for the most part, and cut in several places by deep chasms where the storm surf must create a magnificent scene. Even the peaceful high tides explore them and leave a watermark of rockweeds, barnacles and periwinkles. There is one unexpected, tiny beach where the shore makes a sharp curve and there is a protective jutting out of rocks. At another place, something about the angle of the shore and the set of the currents must have produced just the right conditions to trap the driftwood that comes down the bay, and there is an exciting jumble of logs and treetrunks and stumps of fantastic shape. I suppose there is about a mile of shoreline. Behind this is the wonderful, deep, dark woodland — a cathedral of stillness and peace. Spruce and fir, some hemlock, some pine, and hardwoods along the edges where a fire once destroyed what was there and set in action the restorative forces of nature. It is a living museum of mosses and lichens, which in some places form a carpet many inches deep. Rocks jut out here and there, as a flat floor where only lichens may grow, or rising in shadowed walls. For the most part the woods are dark and silent, but here and there one comes out into open areas of sunshine filled with woods smells. It is a treasure of a place to which I have lost my heart, completely. [. . .]

I have had many precious moments in these woods, and this past fall as I walked there the feeling became overwhelming that something must be done. I had just played a small part in helping to organize a Maine chapter of the Nature Conservancy. My rather nebulous plans of last fall had to do with trying to enlist aid from that quarter. But while the Conservancy can help, the real job has to be pretty well provided for. [. . .]

ᄽ **23** ᄿ

[1957]

Clouds

Television was a new medium for writers in the 1950s, and Carson was not initially enamored of its creative merits. She did, however, recognize television's potential educational value.

When the idea for a show on clouds was proposed by an eight-year-old viewer of the CBS show Omnibus *who wanted to see a program on "something about the sky," the Ford Foundation's TV-Radio Workshop approached Carson to write a television script on clouds. She agreed to collaborate with the* Omnibus *producer and with meteorologist Vincent Schaefer, who had discovered the process of cloud seeding and whose film footage formed the cinematic basis of the show. Her objective was to change the popular conception that cloud types and formations had no particular scientific significance, and to pro-*

vide an awareness of a dynamic process that linked clouds to the broader web of life. The resulting script was vintage Carson, with an emphasis on the long journey of wind and water in a constantly renewing and unending cycle. This venture deeper into the science of weather and climate renewed her interest in writing on the subject of global climate change.

"Something about the Sky" aired on CBS Omnibus *on March 11, 1957, and Carson and her family gathered around her brother's television set to watch her first successful endeavor in an unfamiliar medium. Several days later, Carson capitulated and bought her own television set.*

I. Introduction

(Clouds drifting by, of various types,
but all in motion)

Among the earliest memories of each of us are the images of clouds drifting by overhead, fleecy, fair-weather clouds promising sunny skies — storm clouds bringing portents of rain or snow.

The farmer plowing his field reads the weather language of the sky.

So does the fisherman at sea, and all others who live openly on the face of the earth.

In those of us who live in cities, awareness of the clouds has perhaps grown dim; and even those who live in open country may think of them only as a beautiful backdrop for a rural scene, or an ominous reminder to carry an umbrella today.

The clouds are as old as the earth itself — as much a part of our world as land or sea.
They are the writing of the wind on the sky.
They carry the signature of the masses of air advancing toward us,
across sea or land.
They are the aviator's promise of good flying weather, or an omen of turbulent air.
Most of all they are the cosmic symbols of a process without which life itself could not exist on earth.

II. The Ocean of Air

Today we are going to look at clouds as perhaps we have never looked at them before.
We are going to pretend we live on the bottom of an ocean — an ocean of air in which clouds are adrift — just as sponges and coral and spidery crabs inhabit the floor of the water ocean.

But it will not be hard to pretend that, for in fact that is just what we do. In relation to the air ocean, we are exactly like deep-sea fishes, with all the weight of tons of air pressing down upon our bodies.

And there are other similarities.

Our world is divided into three parts: earth, sea, and air.

Out there is the ocean of water — familiar, though always mysterious.

Its greatest depths lie 7 miles down.

From surface to bottom pressure increases, from 35 pounds to the square inch at the surface to 7 1/2 tons in the greatest depths.

Waves move across it. Great currents flow through it like rivers.

Up there is another ocean — the air ocean that envelops the whole globe.

Its depth, from airless space down to where it touches earth, is some 600 miles.

Like the water ocean, its substance becomes more dense from surface to bottom. Only the lowermost layers are dense enough to support life.

Living on the bottom of this ocean of air, we support on our bodies a pressure of about a ton to the square foot of surface.

In this lower layer, too, clouds are born and die.

Like the sea, the atmospheric ocean is a place of movement and turbulence, stirred by the movements of gigantic waves — torn by the swift passage of winds that are like ocean currents.

Now that we are learning to read the language of the sky, we can interpret much of the structure of our air ocean by looking at the pattern of the clouds.

Look, for instance, at this ribbed pattern of high clouds.

Remember that they are perhaps 8 or 10 miles above us, and so the cloud bands that look rather close together are in reality perhaps 20 miles apart.

Like white caps on the crests of ocean waves, these clouds mark the crests of gigantic atmospheric waves — waves surging through space in an undulating pattern.

The bands of cloud mark the upsurges of condensation; the wave troughs of blue sky, the warmer air valleys of evaporation.

Clouds give other clues to the unseen structure of the ocean of air.

Aviators know the danger of flying in mountainous country, where savage downdrafts in the lee of high peaks may suddenly snatch a plane out of the sky.

The science of clouds is now showing how

271

warning signals are hung out in the sky for all to read them.

When strong winds strike a mountain, the atmosphere up to a height of thousands of feet above the land is thrown into a strong wave motion which extends out over miles of valley on the lee side.

Clouds form on the crests of these atmospheric waves — the strange, almond-shaped lenticular cloud.

So mobile it seems a living thing, the cloud nevertheless maintains its fixed position at the crest of the air waves, neatly balancing condensation and evaporation, built up on the windward side and eaten away on the lee side.

Lenticular clouds scattered out over the valley on the lee side of a mountain range are signs to the pilot of dangerous turbulence.

Here again the clouds are writing a story of violent movement within the atmosphere.

Winds that rush down mountain slopes are known the world over — the foehn wind of the Alps — the chinook of the Rockies — the zonda of the Andes.

As moist air is carried up and over a mountain peak by strong winds, cloud is formed and pours over the crest of the mountain like a waterfall — the foehn wall or foehn cloud.

Here again the flier who can read the clouds stays clear.

These are among the spectacular features of the aerial drama — to which we shall later return. But meanwhile, what of the basic meaning of the clouds — what is their role in the life of the earth?

For us, as living creatures, they are one of the reasons we are men instead of fishes. As land creatures, we must have water.

Without clouds, all water would remain forever in the sea, from which our early ancestors emerged 300 million years ago.

Without the miracle of clouds and rain, the continents would have remained barren and uninhabited, and perhaps life would never have evolved beyond the fishes.

III. The Water Cycle

Almost all of the earth's water is contained in the oceans that encircle the globe — all but a mere three percent.

But to us, inhabitants of the land, that

three percent is vital.

It is engaged in a never-ending cycle of exchange: from sea to air — from air to earth — from earth to sea.

Water from the sea is constantly being brought to the land.

There it makes possible the existence of plants and animals.

There, in streams and rivers, it carves and molds the face of the land, cutting valleys, wearing away hills.

Over all the vast surfaces of the ocean, stirred and broken by the wind, molecules of water vapor are escaping into the overlying air.

This occurs everywhere to some extent, but in the warm tropical seas on each side of the Equator — the belt where the Trade Winds blow — the escape of water vapor into the air is tremendous.

The warm, moist air rises; in the cooler air aloft it condenses.

Processions of woolly cumulus clouds are set drift in the trade wind.

The moisture in these clouds may fall as rain and be recondensed several times, but it eventually becomes part of the vast circulation of the upper atmosphere — drifting over the continents — embodied in the clouds that day after day move from horizon to horizon.

Then in a drama of turbulence and change hidden within the heart of the clouds, the water vapor begins to return to the liquid state — begins to drop earthward with increasing momentum.

Rain falls on the earth — the end of a long journey that began in a tropical sea; yet in a constantly renewing cycle there is no end, as there is no beginning.

Stage succeeds stage, turning again and again upon itself like a wheel.

Or, in the cold regions, snow — a deep, soft, sound-absorbing blanket bringing a great quiet to the earth; storing moisture that will be released gradually to the thirsty land.

From the run-off of high ground — from melting snowfields and glaciers, the water finds its way to the streams: the noisy hill streams tumbling over rocky beds — the quietly rolling waters of the valleys and plains — all to return at last to the sea.

Sometimes the process is marked by the violence of storms

sometimes Nature indulges in the wild fury of floods —

But often the cycle brings us nothing more

troublesome than a gentle April rain —
and always it is in the main a beneficent
process, bringing the continents to life.

IV. Cloud Forms

What of the clouds themselves — the aerial
agents of this cosmic process?
Someone has said that without the gift of
sight, one could never imagine clouds —
their beauty, their ever-changing shapes,
their infinite variety of form.

STRATUS

Rolling, swirling along the floor of the air
ocean are the lowest clouds of all — fog.
For fog is nothing but a stratus cloud so near
the earth that sometimes it touches it.
Fog may shut down quickly over a clear autumn night when the air over the land loses
its heat by rapid evaporation into the open
sky.
Such a fog is a shallow one; though we
earth-bound mortals grope blindly through
it, the tops of tall trees may clear it, and in
the morning the sun quickly burns it
away.
Fog of a different sort forms when warm sea
air rolls in over colder coastal waters and

276

over the land — shutting down harbors — grounding planes — isolating ships at sea with its soft grey swirling mists.

When a fog drifts at a height of a thousand feet or so, forming the aviator's "ceiling," it is really a layer cloud called stratus.

As we fly above it, it is a veil through which the earth is seen dimly, like the shallow bottom of a bay when one looks down from an idling skiff.

Or it may stretch away to the plane's horizon like a monotonous Arctic icefield.

Compared with the high-drifting cirrus wisps and the soaring columns of the cumulus, the stratus clouds are the duller earthlings — coarse-textured clouds formed of large water droplets.

CUMULUS

Most beautiful in the infinite variety of their shapes are the cumulus clouds.

These are also the clouds that generate the most incredible violence known on earth — for beside the power of a tornado or a hurricane even the atomic bomb is insignificant.

The birth of a cumulus cloud is relatively peaceful and simple.

As the earth warms under the morning

sun, it heats unevenly.

Invisible columns of warm air begin to rise — from a plowed field, a lake, a town — any area warmer than its surroundings.

The column of rising air contains invisible molecules of water vapor drawn from vegetation, evaporated from the surface of earth or water.

Such warm air can hold quantities of water in the vapor state.

Rising, it cools; at a certain point it can no longer contain its water invisibly; and the white misty substance of a cloud is born.

Broad-winged birds like hawks and eagles find these soaring "thermals" and ride them for hours.

Glider pilots seek them out, locating them by the clouds at their summits.

Polynesian navigators steering across the South Pacific from atoll to atoll, find their way by the cloud rising like a kite from each pinpoint of warm land.

Most cumulus clouds have straight-edge bases, as though evened off by the stroke of a cosmic knife — but the shaping blade is the altitude that marks a sharp change to cooler temperature — below this line the air column holds its vapor invisibly — once above it, all the water molecules blossom, through condensation, into the fabric of a cloud.

In regions of very warm, moist air, the atmosphere is in the power of highly unstable forces.

Then a cumulus cloud puffs up and up to extraordinary heights.

When the tornado of June 9th, 1953, approached the city of Worcester, Mass., observers at MIT reported that the clouds towered right off the radar screens, which could register only to altitudes of 50,000 feet.

Even higher clouds are known from the true tornado country of the middle west — 70,000-foot giants more than twice as high as Everest.

CIRRUS

Most ethereal and fragile of all are the high-floating wisps of cirrus, drifting just under the stratosphere.

If we could approach them closely in an airplane we would find them glittering in iridescent splendor like the dust of diamonds.

Up in these substratospheric vaults of sky, from which the earth looks like the sphere it is, there is a hard, bitter cold, far below zero, summer and winter.

So the cirrus clouds are composed of minute crystals of ice — the merest specks of substance, so thinly spread through the sky that not more than 2 or 3 occupy a cubic inch of space.

It is the high-riding cirrus that first beholds the sunrise, or in evening holds the light of sunset longest, reflecting back to the dark earth the splendor of a light no longer visible — the rose and gold, the wine and scarlet of the sun.

The cirrus clouds are the birthplace of the snow, slowly cascading down to earth in long, curving streaks as the crystals fall behind the swift winds of the upper sky.

Halos seen around the sun or moon are the ice crystals of a cirrus veil called cirrostratus.

Like the lower clouds, cirrus is formed of water vapor that is drawn from the sea and pumped aloft in the swift updrafts of cumulus clouds — or that rides up an ascending elevator of warm air slipping over a cold front.

Sometimes cirrus is born of material torn from the top of high cumulus by the strong winds of the upper air — winds that tear off the crests of the clouds as, at sea, a gale blows off the wave crests and sends the

spindrift scudding away over the water.
Sweeping curls of cirrus indicate the passage overhead of a rushing current of air, pouring through the sky at a speed of two or three hundred miles an hour.

[ed.: Carson ends the script with the story of jet streams, the strongest of all winds, and conjectures that the forces that direct the jet stream will be "found in the far depths of the sky [and] written in the clouds."]

Part Four

Part Four covers the period 1959–1963. During that time Carson was occupied with either writing or defending Silent Spring, which she had initially titled "The Control of Nature" when she began her research in the fall of 1957. It took nearly five years for her to gather evidence, synthesize, and shape the enormous body of scientific literature into a compelling indictment against the flagrant misuse of synthetic chemical pesticides, and the folly of trying to conquer nature.

Included in Part Four are three of Carson's most important public speeches, as notable for their clarity of language as for their expression of her convictions about both the dangers of pollution and the interconnectedness of life. Attacks on Carson and her work increased after 1962, and she answered her critics with a calm but compelling analysis and unexpected political insight. Carson had attacked the integrity of the scientific *establishment, its moral leadership, and its direction of society. She exposed their self-interest as well as their poor science, and defended the public's right to know the truth.*

At the same time as Carson carried out her public crusade she was fighting an even graver private adversary. Diagnosed with an aggressively metastasizing breast cancer in 1961, she defended the earth she loved with an added passion born of knowing that her opportunities to speak out were now limited. Part Four ends with letters to her physician and to her dearest friend.

∿ 24 ∿

[1959]

Vanishing Americans

*Carson had been working on the book that be-
came* Silent Spring *for nearly two years when
the* Washington Post *published an editorial
commenting on a recent National Audubon Society report describing the effects of an unusually
harsh winter on migrating birds in the South.
Knowing that climate variations explained only
a small part of the population decline, Carson
wrote exposing the role the widespread use of
toxic chemicals played in "silencing the birds."
Her focus on birds offered a good opportunity to
gauge the public's awareness of the pesticide
problem.*

*Her letter, published in the newspaper a week
later, provided the first clue that Rachel Carson
was studying the subject of synthetic pesticides.
She was gratified when the public response to her*

article testified to an intense interest in the subject.

An added benefit of the publication of Carson's letter was the support of the Washington Post *owner Agnes Meyer and of activist Christine Stevens, president of the Animal Welfare Institute in New York. Both women subsequently became influential advocates of Carson's work.*

Your excellent March 30 editorial, "Vanishing Americans," is a timely reminder that in our modern world nothing may be taken for granted — not even the spring songs that herald the return of the birds. Snow, ice and cold, especially when visited upon usually temperate regions, leave destruction behind them, as was clearly brought out in the report of the National Audubon Society you quote.

But although the recent severe winters in the South have taken their toll of bird life, this is not the whole story, nor even the most important part of the story. Such severe winters are by no means rare in the long history of the earth. The natural resilience of birds and other forms of life allows them to take these adverse conditions in their stride and so to recover from temporary reduction of their populations.

It is not so with the second factor, of which

you make passing mention — the spraying of poisonous insecticides and herbicides. Unlike climatic variations, spraying is now a continuing and unremitting factor.

During the past 15 years, the use of highly poisonous hydrocarbons and of organic phosphates allied to the nerve gases of chemical warfare has built up from small beginnings to what a noted British ecologist recently called "an amazing rain of death upon the surface of the earth." Most of these chemicals leave long-persisting residues on vegetation, in soils, and even in the bodies of earthworms and other organisms on which birds depend for food.

The key to the decimation of the robins, which in some parts of the country already amounts to virtual extinction, is their reliance on earthworms as food. The sprayed leaves with their load of poison eventually fall to become part of the leaf litter of the soil; earthworms acquire and store the poisons through feeding on the leaves; the following spring the returning robins feed on the worms. As few as 11 such earthworms are a lethal dose, a fact confirmed by careful research in Illinois.

The death of the robins is not mere speculation. The leading authority on this problem, Professor George Wallace of Michigan State University, has recently reported that "Dead and dying robins, the latter most often found in a state of violent convulsions, are most common in the spring, when warm rains bring up the earthworms, but birds that survive are apparently sterile or at least experience nearly complete reproductive failure."

The fact that doses that are sub-lethal may yet induce sterility is one of the most alarming aspects of the problem of insecticides. The evidence on this point, from many highly competent scientists, is too strong to question. It should be weighed by all who use the modern insecticides, or condone their use.

I do not wish to leave the impression that only birds that feed on earthworms are endangered. To quote Professor Wallace briefly: "Tree-top feeders are affected in an entirely different way, by insect shortages, or actual consumption of poisoned insects. [. . .] Birds that forage on trunks and branches are also affected, perhaps mostly by the dormant sprays." About two-thirds of the bird species that were formerly summer residents in the area under Professor Wallace's observation have disappeared entirely or are sharply reduced.

To many of us, this sudden silencing of the song of birds, this obliteration of the color and beauty and interest of bird life, is sufficient cause for sharp regret. To those who have never known such rewarding enjoyment of nature, there should yet remain a nagging and insistent question: If this "rain of death" has produced so disastrous an effect on birds, what of other lives, including our own?

✌ 25 ✌

[1960, 1964]

To Understand Biology/
Preface to *Animal Machines*

Carson agreed to contribute an introduction for the Animal Welfare Institute's educational book-let, Humane Biology Projects, *which addressed the need for reform of biology instruction in the nation's high schools. The Institute opposed animal experimentation and worked to change the callous attitude toward systematic cruelty that often accompanied classroom biology projects.*

A.W.I. head Christine Stevens was also instrumental in introducing Carson to the work of British activist Ruth Harrison, whose book Animal Machines *exposed the inhumane methods of raising livestock and the deplorable conditions in which they were kept before slaughter. In 1963 Carson wrote the preface to Harrison's book.*

Carson's ideas about the humane treatment of animals place her fully in the tradition of Albert Schweitzer and his philosophy of the reverence for life. Her contributions to these publications emphasize the unity of all life, and the need to cultivate an emotional response to the living world.

Through the next several years Carson quietly aided the work of Stevens and the Animal Welfare Institute, writing to members of Congress in support of legislation banning the use of certain leg traps and against the inhumane treatment of laboratory animals. But she had to be careful not to draw too much attention to her support for causes that might link her in the public mind with fringe groups and extremists, lest she jeopardize her all-important work concerning the misuse of pesticides. Had this not been a real political consideration, Carson undoubtedly would have been an outspoken advocate of the humane treatment of animals.

To Understand Biology

I like to define biology as the history of the earth and all its life — past, present, and future. To understand biology is to understand that all life is linked to the earth from which it came; it is to understand that the stream of

291

life, flowing out of the dim past into the uncertain future, is in reality a unified force, though composed of an infinite number and variety of separate lives. The essence of life is lived in freedom. Any concept of biology is not only sterile and profitless, it is distorted and untrue, if it puts its primary focus on unnatural conditions rather than on those vast forces not of man's making that shape and channel the nature and direction of life.

To the extent that it is ever necessary to put certain questions to nature by placing unnatural restraints upon living creatures or by subjecting them to unnatural conditions or to changes in their bodily structure, this is a task for the mature scientist. It is essential that the beginning student should first become acquainted with the true meaning of his subject through observing the lives of creatures in their true relation to each other and to their environment. To begin by asking him to observe artificial conditions is to create in his mind distorted conceptions and to thwart the development of his natural emotional response to the mysteries of the life stream of which he is a part. Only as a child's awareness and reverence for the wholeness of life are developed can his humanity to his own kind reach its full development.

Preface to *Animal Machines*

The modern world worships the gods of speed and quantity, and of the quick and easy profit, and out of this idolatry monstrous evils have arisen. Yet the evils go long unrecognized. Even those who create them manage by some devious rationalizing to blind themselves to the harm they have done society. As for the general public, the vast majority rest secure in a childlike faith that "someone" is looking after things — a faith unbroken until some public-spirited person, with patient scholarship and steadfast courage, presents facts that can no longer be ignored.

This is what Ruth Harrison has done. Her theme affects practically every citizen, for it deals with the new methods of rearing animals destined to become human food. It is a story that ought to shock the complacency out of any reader.

Modern animal husbandry has been swept by a passion for "intensivism"; on this tide everything that resembles the methods of an earlier day has been carried away. Gone are the pastoral scenes in which animals wandered through green fields or flocks of chickens scratching contentedly for their food. In their place are factorylike buildings in which animals live out their wretched existences without ever feeling the earth beneath their feet, without knowing sunlight, or experiencing the simple pleasures of grazing for natural food — indeed, so confined or so intolerably crowded that movement of any kind is scarcely possible. [. . .]

As a biologist whose special interests lie in the field of ecology, or the relation between living things and their environment, I find it inconceivable that healthy animals can be produced under the artificial and damaging conditions that prevail in these modern and factorylike installations, where animals are grown and turned out like so many inanimate objects. The intolerable crowding of broiler chickens, the revoltingly unsanitary conditions in the piggeries, the lifelong confinement of laying hens in tiny cages are samples of the conditions Mrs. Harrison describes. As she makes abundantly clear, this artificial environment is not a healthy one. Diseases sweep through these establishments, which indeed are kept going only by the continuous administration of antibiotics. Disease organisms then become resistant to the antibiotics. Veal calves, purposely kept in a state of induced aenemia so their white flesh will satisfy the supposed desires of the gourmet, sometimes drop dead when taken out of their imprisoning crates.

The question then arises: how can animals produced under such conditions be safe or acceptable human food? Mrs. Harrison quotes expert opinion and cites impressive evidence that they are not. Although the quantity of production is up, quality is down, a fact recognized in a most significant way by some of the producers themselves, who, for example, are more likely to keep a few chickens in the back yard for their own tables than to eat the products of the broiler establishments. The menace to human consumers from the drugs, hormones, and pesticides used to keep this whole fantastic operation somehow going is a matter never properly explored.

The final argument against the intensivism now practiced in this branch of agriculture is a humanitarian one. I am glad to see Mrs. Harrison raise the question of how far man has a moral right to go in his domination of other life. Has he the right, as in these examples, to reduce life to a bare existence that is scarcely life at all? Has he the further right to terminate these wretched lives by means that are wantonly cruel? My own answer is an unqualified no. It is my belief that man will never be at peace with his own kind until he has recognized the Schweitzerian ethic that embraces

decent consideration for all living creatures —
a true reverence for life.

Although Mrs. Harrison's book describes
in detail only the conditions prevailing in
Great Britain, it deserves to be widely read
also in those European countries where these
methods are practiced, and in the United
States where some of them arose. Wherever it
is read it will certainly provoke feelings of
dismay, revulsion, and outrage. I hope it will
spark a consumers' revolt of such proportions
that this vast new agricultural industry will be
forced to mend its ways.

~: 26 :~

[1962]

A Fable for Tomorrow

The brief fable with which Carson opens Silent Spring *is one of the most memorable in contemporary nonfiction and elicited more controversy than almost any other part of the book. Many scientists were appalled that Carson dared begin a book about the science of chemical pesticides with an allegory about the environmental pollution of an imaginary town. Some simply ignored the fact that it was a fable and attacked Carson because the town was not accurately described, while others accused her of writing science fiction throughout. By contrast, most literary critics praised her use of the fable as a brilliant rhetorical device and a creative way of introducing the disturbing subject of the deliberate poisoning of the earth.*

Carson realized her first chapter, originally ti-

297

tled "The Rain of Death," might be too formidable and used the fable as a device to engage the nonscientific reader. In early drafts, Carson gave her town a name, Green Meadows, and centered the action on a young man who returns home after many years only to find his town devastated by ecological havoc. At the urging of her publisher, she rewrote the fable making it clear that the town was a composite of many communities and became the voice of the fable's narrator. The opening paragraphs recall the once bucolic town of Springdale, Pennsylvania, where Carson grew up, which was subjected to an earlier kind of industrial pollution.

There was once a town in the heart of America where all life seemed to live in harmony with its surroundings. The town lay in the midst of a checkerboard of prosperous farms, with fields of grain and hillsides of orchards where, in spring, white clouds of bloom drifted above the green fields. In autumn, oak and maple and birch set up a blaze of color that flamed and flickered across a backdrop of pines. Then foxes barked in the hills and deer silently crossed the fields, half hidden in the mists of the fall mornings.

Along the roads, laurel, viburnum and alder, great ferns and wildflowers delighted the traveler's eye through much of the year. Even in winter the roadsides were places of beauty, where countless birds came to feed on the berries and on the seed heads of the dried weeds rising above the snow. The countryside was, in fact, famous for the abundance and variety of its bird life, and when the flood of migrants was pouring through in spring and fall people traveled from great distances to observe them. Others came to fish the streams, which flowed clear and cold out of the hills and contained shady pools where trout lay. So it had been from the days many years ago when the first settlers raised their houses, sank their wells, and built their barns.

Then a strange blight crept over the area and everything began to change. Some evil spell had settled on the community: mysterious maladies swept the flocks of chickens; the cattle and sheep sickened and died. Everywhere was a shadow of death. The farmers spoke of much illness among their families. In the town the doctors had become more and more puzzled by new kinds of sickness appearing among their patients. There had been several sudden and unexplained deaths, not only among adults but even among children, who would be stricken suddenly while at play and die within a few hours.

There was a strange stillness. The birds, for

example — where had they gone? Many people spoke of them, puzzled and disturbed. The feeding stations in the backyards were deserted. The few birds seen anywhere were moribund; they trembled violently and could not fly. It was a spring without voices. On the mornings that had once throbbed with the dawn chorus of robins, catbirds, doves, jays, wrens, and scores of other bird voices there was now no sound; only silence lay over the fields and woods and marsh.

On the farms the hens brooded, but no chicks hatched. The farmers complained that they were unable to raise any pigs — the litters were small and the young survived only a few days. The apple trees were coming into bloom but no bees droned among the blossoms, so there was no pollination and there would be no fruit.

The roadsides, once so attractive, were now lined with browned and withered vegetation as though swept by fire. These, too, were silent, deserted by all living things. Even the streams were now lifeless. Anglers no longer visited them, for all the fish had died.

In the gutters under the eaves and between the shingles of the roofs, a white granular powder still showed a few patches; some weeks before it had fallen like snow upon the roofs and the lawns, the fields and streams.

No witchcraft, no enemy action had silenced the rebirth of new life in this stricken world. The people had done it themselves.

This town does not actually exist, but it might easily have a thousand counterparts in America or elsewhere in the world. I know of no community that has experienced all the misfortunes I describe. Yet every one of these disasters has actually happened somewhere, and many real communities have already suffered a substantial number of them. A grim specter has crept upon us almost unnoticed, and this imagined tragedy may easily become a stark reality we all shall know.

What has already silenced the voices of spring in countless towns in America? This book is an attempt to explain.

Women's National Press Club Speech

Silent Spring *was serialized in three summer is-*
sues of the New Yorker *in 1962 and published*
in late September. The high level of public inter-
est that surrounded the book included notice by
President John F. Kennedy, who convened a
special panel of the President's Science Advisory
Committee to look into the problem, the introduc-
tion of legislation in several states seeking to halt
the spraying of pesticides without citizen notifica-
tion, and general uproar in the agricultural
chemical industry and among government scien-
tists.

Carson took many of her critics in stride, but
she could not abide those that damned the book
without having read it. As debate grew more ac-
rimonious in the fall of 1962, Carson's public re-
marks grew sharper, culminating in her

appearance at the Women's National Press Club in December. In this speech, Carson attacked the smugly self-satisfied chemical industry and exposed their counterparts in industry-funded research institutions.

With national television cameras rolling, Carson charged that basic scientific truths were being compromised "to serve the gods of profit and production."

My text this afternoon is taken from the *Globe Times* of Bethlehem, Pa., a news item in the issue of October 12. After describing in detail the adverse reactions to *Silent Spring* of the farm bureaus in two Pennsylvania counties, the reporter continued: "No one in either county farm office who was talked to today had read the book, but all disapproved of it heartily."

This sums up very neatly the background of much of the noisier comment that has been heard in this unquiet autumn following the publication of *Silent Spring*. In the words of an editorial in the *Bennington Banner*, "The anguished reaction to *Silent Spring* has been to refute statements that were never made." Whether this kind of refutation comes from people who actually have not read the book or from those who find it convenient to misrepresent my position I leave it to others to judge.

Early in the summer — as soon as the first installment of the book appeared in the *New Yorker* — public reaction to *Silent Spring* was reflected in a tidal wave of letters — letters to Congressmen, to newspapers, to Government agencies, to the author. These letters continue to come and I am sure represent the most important and lasting reaction.

Even before the book was published, editorials and columns by the hundred had discussed it all over the country. Early reaction in the chemical press was somewhat moderate, and in fact I have had fine support from some segments of both chemical and agricultural press. But in general, as was to be expected, the industry press was not happy. By late summer the printing presses of the pesticide industry and their trade associations had begun to pour out the first of a growing stream of booklets designed to protect and repair the somewhat battered image of pesticides. Plans are announced for quarterly mailings to opinion leaders and for monthly news stories to newspapers, magazines, radio, and television. Speakers are addressing audiences everywhere.

It is clear that we are all to receive heavy doses of tranquilizing information, designed to lull the public into the sleep from which *Silent Spring* so rudely awakened it. Some definite gains toward a saner policy of pest control have been made in recent months. The important issue now is whether we are to hold and extend those gains.

The attack is now falling into a definite pattern and all the well-known devices are being used. One obvious way to try to weaken a cause is to discredit the person who champions it. So the masters of invective and insinuation have been busy: I am a "bird lover — a cat lover — a fish lover" — a priestess of nature — a devotee of a mystical cult having to do with laws of the universe which my critics consider themselves immune to.

Another piece in the pattern of attack largely ignores *Silent Spring* and concentrates on what I suppose would be called the soft sell, the soothing reassurances to the public. Some of these acknowledge the correctness of my facts, but say that the incidents I reported occurred some time in the past, that industry and Government are well aware of them and have long since taken steps to prevent their recurrence. It must be assumed that the people who read these comforting reports read nothing else in their newspapers. Actually, pesticides have figured rather prominently in the news in recent months: some items trivial, some almost humorous, some definitely serious.

These reports do not differ in any important way from the examples I cited in *Silent Spring*, so if the situation is under better control there is little evidence of it.

What are some of the ways pesticides have made recent news?

1. The *New York Post* of October 12 reported the seizure by the Food and Drug Administration of more than a quarter of a million pounds of potatoes — 346,000 pounds to be exact — in the Pacific Northwest. Agents said they contained about 4 times the permitted residues of aldrin and dieldrin.

2. In September, Federal investigators had to look into the charge that vineyards near the Erie County thruway had been damaged by weed-killer chemicals sprayed along the highway. Similar reports came from Iowa.

3. In California, fumes from lawns to which a chemical had been applied were so obnoxious that the fire department was called to drench the lawns with water. Thereupon the fumes increased so greatly that 11 firemen were hospitalized.

4. Last summer the newspapers widely reported the story of some 5000 Turkish children suffering from an affliction called *porphyria* characterized by severe liver damage and the growth of hair on face, hands, and arms, giving a monkey-like appearance to victims. This was traced to the consumption of wheat treated with a chemical fungicide. The wheat had been intended for planting, rather than for direct consumption. But the people were hungry and perhaps did not understand the restriction. This was an unplanned occurrence in a far part of the world but it is well to remember that large quantities of seed are similarly treated here.

5. You will remember that the bald eagle, our national emblem, is seriously declining in numbers. The Fish and Wildlife Service recently reported significant facts that may explain why this is so. The Service has determined experimentally how much DDT is required to kill an eagle. It has also discovered that eagles found dead in the wild have lethal doses of DDT stored in their tissues.

6. This fall also, Canadian papers carried a warning that woodcock being shot during the hunting season in New Brunswick were carrying residues of heptachlor and might be dangerous if used as food. Woodcock are migratory birds. Those that nest in New Brunswick winter in the southern United States,

where heptachlor has been used extensively in the campaign against the fire ant. The residues in the birds were 3 to 3.5 ppm. The legal tolerance for heptachlor is ZERO.

7. Biologists of the Massachusetts Fish and Game Department have recently reported that fish in the Framingham Reservoir on the outskirts of Boston contain DDT in amounts as high as 75 ppm, or more than 10 times the legal tolerance. This is, of course, a public water supply for a large number of people.

8. One more item — an Associated Press dispatch of November 16th: a sad commentary on technology gone wrong. A Federal Court Jury awarded a New York State farmer *$12,360* for damages to his potato crop. The damage was done by a chemical that was supposed to halt sprouting. Instead, the sprouts grew inward.

We are told also that chemicals are never used unless tests have shown them to be safe. This, of course, is not an accurate statement. I am happy to see that the Department of Agriculture plans to ask the Congress to amend the FIFRA to do away with the provision that now permits a company to register a pesticide under protest, even though a question of health or safety has been raised by the Department.

We have other reminders that unsafe chemicals get into use — County Agents frequently have to amend or rescind earlier advices on the use of pesticides. For example, a letter was recently sent out to farmers recalling stocks of a chemical in use as a cattle spray. In September, "unexplained losses" occurred following its use. Several suspected production lots were recalled but the losses continued. All outstanding lots of the chemical have now had to be recalled.

Inaccurate statements in reviews of *Silent Spring* are a dime a dozen, and I shall only mention one or two examples. *Time*, in its discussion of *Silent Spring*, described accidental poisonings from pesticides as *very rare*. Let's look at a few figures. California, the only state that keeps accurate and complete records, reports from 900 to 1000 cases of poisoning from agricultural chemicals per year. About 200 of these are from parathion alone. Florida has experienced so many poisonings recently that this state has attempted to control the use of the more dangerous chemicals in residential areas. As a sample of conditions in other countries, parathion was responsible for 100 deaths in India in 1958 and takes an average of 336 deaths a year in Japan.

It is also worthy of note that during the years 1959, 1960, and 1961, airplane crashes involving crop-dusting planes totaled 873. In these accidents 135 pilots lost their lives. This very fact has led to some significant research by the Federal Aviation Agency through its Civil Aeromedical Unit — research designed to find out *why* so many of these planes crashed. These medical investigators took as their basic premise the assumption that spray poisons accumulate in the pilot's body — inside the cells, where they are difficult to detect.

These researchers recently reported that they had confirmed two very significant facts: 1. That there is a causal relation between the build-up of toxins in the cell and the onset of sugar diabetes. 2. That the build-up of poisons within the cell interferes with the rate of energy production in the human body.

I am, of course, happy to have this confirmation that cellular processes are not so "irrelevant" as a certain scientific reviewer of *Silent Spring* has declared them to be.

This same reviewer, writing in a chemical journal, was much annoyed with me for giving the sources of my information. To identify the person whose views you are quoting is, according to this reviewer, *name-dropping*. Well, times have certainly changed since I received my training in the scientific method at Johns Hopkins! My critic also *profoundly disapproved* of my bibliography. The very fact that it gave complete and specific references for each important statement was extremely distasteful to him. This was *padding* to impress the uninitiated with its length.

Now I would like to say that in *Silent Spring* I have never asked the reader to take *my* word. I have given him a very clear indication of my sources. I make it possible for him — indeed I invite him — to go beyond what I report and get the full picture. This is the reason for the 55 pages of references. You cannot do this if you are trying to conceal or distort or to present half truths.

Another reviewer was offended because I made the statement that it is customary for pesticide manufacturers to support research on chemicals in the universities. Now, this is just common knowledge and I can scarcely believe the reviewer is unaware of it, because his own university is among those receiving such grants.

But since my statement has been challenged, I suggest that any of you who are interested make a few inquiries from representative universities. I am sure you will find out that the practice is very widespread. Actually, a visit to a good scientific library will quickly establish the fact, for it is still generally the custom for authors of technical papers to acknowledge the source of funds for the investigation. For example, a few gleaned at random from the *Journal of Economic Entomology* are as follows:

1. In a paper from Kansas State University, a footnote states: Partial cost of publication of this paper was met by the Chemagro Corporation.
2. From the University of California Citrus Experimental Station: The authors thank the Diamond Black-Leaf Co., Richmond, Virginia, for grants-in-aid.
3. University of Wisconsin: Research was also supported in part by grants from the Shell Chemical Co., Velsicol Chemical Corporation and Wisconsin Canners Association.
4. Illinois Nat. Hist. Survey: This investigation was sponsored by the Monsanto Chem. Co. of St. Louis, Mo.

A penetrating observer of social problems has pointed out recently that whereas wealthy families once were the chief benefactors of the Universities, now industry has taken over this role. Support of education is something no one quarrels with — but this need not blind us to the fact that research supported by pesticide manufacturers is not likely to be directed at discovering facts indicating unfavorable effects of pesticides.

Such a liaison between science and industry is a growing phenomenon, seen in other areas as well. The AMA, through its newspaper, has just referred physicians to a pesticide trade association for information to help them answer patients' questions about the effects of pesticides on man. I am sure physicians have a need for information on this subject. But I would like to see them referred to authoritative scientific or medical literature — not to a trade organization whose business it is to promote the sale of pesticides.

We see scientific societies acknowledging as "sustaining associates" a dozen or more giants of a related industry. When the scientific organization speaks, whose voice do we hear — that of science? or of the sustaining industry? It might be a less serious situation if this voice were always clearly identified, but the public assumes it is hearing the voice of science.

What does it mean when we see a committee set up to make a supposedly impartial review of a situation, and then discover that the committee is affiliated with the very industry whose profits are at stake? I have this week read two reviews of the recent reports of a National Academy of Sciences Committee on the relations of pesticides to wildlife. These reviews raise disturbing questions. It is important to understand just what this committee is. The two sections of its report that have now been published are frequently cited by the pesticide industry in attempts to refute my statements. The public, I believe, assumes that the Committee is actually part of the Academy. Although appointed by the Academy, its members come from outside. Some are scientists of distinction in their fields. One would suppose the way to get an impartial evaluation of the impact of pesticides on wildlife would be to set up a committee of completely disinterested individuals. But the review appearing this week in *The Atlantic Naturalist* described the composition of the Committee as follows: "A very significant role in this committee is played by the Liaison Representatives. These are of three categories. A.) Supporting Agencies. B.) Government Agencies. C.) Scientific Societies. The supporting agencies are presumably those who supply the hard cash. Forty-three such agencies are listed, including 19 chemical companies comprising the massed might of the chemical industry. In addition, there are at least 4 trade organizations such as the National Agricultural Chemical Association and the National Aviation Trades Association."

The Committee reports begin with a firm statement in support of the use of chemical pesticides. From this predetermined position, it is not surprising to find it mentioning only *some* damage to *some* wildlife. Since, in the modern manner, there is no documentation, one can neither confirm or deny its findings. *The Atlantic Naturalist* reviewer described the reports as "written in the style of a trained public relations official of industry out to placate some segments of the public that are causing trouble."

All of these things raise the question of the communication of scientific knowledge to the public. Is industry becoming a screen through which facts must be filtered, so that the hard, uncomfortable truths are kept back and only the harmless morsels allowed to filter through? I know that many thoughtful scientists are deeply disturbed that their organizations are becoming *fronts* for industry. More than one scientist has raised a disturbing question — whether a spirit of lysenkoism may be developing in America today — the philosophy that perverted and destroyed the science of genetics in Russia and even infiltrated all of that nation's agricultural sciences.

But here the tailoring, the screening of basic truth, is done, not to suit a party line, but to accommodate to the short-term gain, to serve the gods of profit and production.

These are matters of the most serious importance to society. I commend their study to you, as professionals in the field of communication.

∿ 28 ∿

[1963]

A New Chapter to *Silent Spring*

As Carson learned of further incidents of pesticide damage and injury from other scientists and from the letters she received from readers, she included this new information every time she spoke in public. Her speeches during this last year of her life reflect her moral conviction that "no civilization can wage relentless war on life without destroying itself, and without losing the right to be called civilized."

Her address to the women of the Garden Club of America in January, 1963, opened a new, aggressively political phase of the pesticide struggle. Here Carson focused specifically on the economic and political forces that prevented changes in pesticide policy, and she urged individuals to demand change in their communities, encouraging grassroots activities to reform the system.

She also addressed the stream of propaganda that had issued from pesticide trade groups, misinformation that hid their true links to industry behind bland affiliations to research organizations or educational institutions. The speech reveals Carson as a tough and trenchant political infighter who understood the nature of her opposition, and who wisely directed her message to concerned individuals, such as the activist women of the nation's garden clubs.

I am particularly glad to have this opportunity to speak to you. Ever since, ten years ago, you honored me with your Frances Hutchinson medal, I have felt very close to The Garden Club of America. And I should like to pay tribute to you for the quality of your work and for the aims and aspirations of your organization. Through your interest in plant life, your fostering of beauty, your alignment with constructive conservation causes, you promote that onward flow of life that is the essence of our world.

This is a time when forces of a very different nature too often prevail — forces careless of life or deliberately destructive of it and of the essential web of living relationships.

My particular concern, as you know, is with the reckless use of chemicals so unselective in their action that they should more appropriately be called biocides rather than pesticides. Not even their most partisan defenders can claim that their toxic effect is limited to insects or rodents or weeds or whatever the target may be.

The battle for a sane policy for controlling unwanted species will be a long and difficult one. The publication of *Silent Spring* was neither the beginning nor the end of that struggle. I think, however, that it is moving into a new phase, and I would like to assess with you some of the progress that has been made and take a look at the nature of the struggle that lies before us.

We should be very clear about what our cause is. What do we oppose? What do we stand for? If you read some of my industry-oriented reviewers you will think that I am opposed to any efforts to control insects or other organisms. This, of course, is *not* my position and I am sure it is not that of The Garden Club of America. We differ from the promoters of biocides chiefly in the means we advocate, rather than the end to be attained.

It is my conviction that if we automatically call in the spray planes or reach for the aerosol bomb when we have an insect problem we are resorting to crude methods of a rather low scientific order. We are being particularly unscientific when we fail to press forward with research that will give us the new kind of weapons we need. Some such weapons now exist — brilliant and imaginative prototypes of what I trust will be the insect control methods of the future. But we need many more, and we need to make better use of those we have. Research men of the Department of Agriculture have told me privately that some of the measures they have developed and tested and turned over to the insect control branch have been quietly put on the shelf.

I criticize the present heavy reliance upon biocides on several grounds: First, on the grounds of their inefficiency. I have here some comparative figures on the toll taken of our crops by insects before and after the DDT era. During the first half of this century, crop loss due to insect attack has been estimated by a leading entomologist at 10 percent a year. It is startling to find, then, that the National Academy of Science last year placed the present crop loss at 25 percent a year. If the percentage of crop loss is increasing at this rate, even as the use of modern insecticides increases, surely something is wrong with the methods used! I would remind you that a non-chemical method gave 100 percent control of the screwworm fly — a degree of success no chemical has ever achieved.

Chemical controls are inefficient also because as now used they promote resistance among insects. The number of insect species resistant to one or more groups of insecticides has risen from about a dozen in pre-DDT days to nearly 150 today. This is a very serious problem, threatening, as it does, greatly impaired control.

Another measure of inefficiency is the fact that chemicals often provoke resurgences of the very insect they seek to control, because they have killed off its natural controls. Or they cause some other organism suddenly to rise to nuisance status: spider mites, once relatively innocuous, have become a world-wide pest since the advent of DDT.

My other reasons for believing we must turn to other methods of controlling insects have been set forth in detail in *Silent Spring* and I shall not take time to discuss them now. Obviously, it will take time to revolutionize our methods of insect and weed control to the point where dangerous chemicals are minimized. Meanwhile, there is much that can be done to bring about some immediate improvement in the situation through better procedures and controls.

In looking at the pesticide situation today, the most hopeful sign is an awakening of strong public interest and concern. People are beginning to ask questions and to insist upon proper answers instead of meekly acquiescing in whatever spraying programs are proposed. This in itself is a wholesome thing.

There is increasing demand for better legislative control of pesticides. The state of Massachusetts has already set up a Pesticide Board with actual authority. This Board has taken a very necessary step by requiring the licensing of anyone proposing to carry out aerial spraying. Incredible though it may seem, before this was done anyone who had money to hire an airplane could spray where and when he pleased. I am told that the state of Connecticut is now planning an official investigation of spraying practices. And of course on a national scale, the President last summer directed his science advisor to set up a committee of scientists to review the whole matter of the government's activities in this field.

Citizens groups, too, are becoming active. For example, the Pennsylvania Federation of Women's Clubs recently set up a program to protect the public from the menace of poisons in the environment — a program based on education and promotion of legislation. The National Audubon Society has advocated a 5-point action program involving both state and federal agencies. The North American Wildlife Conference this year will devote an important part of its program to the problem of pesticides. All these developments will serve to keep public attention focused on the problem.

I was amused recently to read a bit of wishful thinking in one of the trade magazines. Industry "can take heart," it said, "from the fact that the main impact of the book (i.e., *Silent Spring*) will occur in the late fall and winter — seasons when consumers are not normally active buyers of insecticides [. . .] it is fairly safe to hope that by March or April *Silent Spring* no longer will be an interesting conversational subject."

If the tone of my mail from readers is any guide, and if the movements that have already been launched gain the expected momentum, this is one prediction that will not come true.

This is not to say that we can afford to be complacent. Although the attitude of the public is showing a refreshing change, there is very little evidence of any reform in spraying practices. Very toxic materials are being applied with solemn official assurances that they will harm neither man nor beast. When wildlife losses are later reported, the same officials deny the evidence or declare the animals must have died from "something else."

Exactly this pattern of events is occurring in a number of areas now. For example, a newspaper in East St. Louis, Illinois, describes the death of several hundred rabbits, quail and songbirds in areas treated with pellets of the insecticide, dieldrin. One area involved was, ironically, a "game preserve." This was part of a program of Japanese beetle control.

The procedures seem to be the same as those I described in *Silent Spring*, referring to another Illinois community, Sheldon. At Sheldon the destruction of many birds and small mammals amounted almost to annihilation. Yet an Illinois Agriculture official is now quoted as saying dieldrin has no serious effect on animal life.

A significant case history is shaping up now in Norfolk, Virginia. The chemical is the very toxic dieldrin, the target the white fringed beetle, which attacks some farm crops. This situation has several especially interesting features. One is the evident desire of the state agriculture officials to carry out the program with as little advance discussion as possible. When the Outdoor Edition of the *Norfolk Virginian-Pilot* "broke" the story, he reported that officials refused comment on their plans. The Norfolk health officer offered reassuring statements to the public on the grounds that the method of application guaranteed safety:

The poison would be injected into the ground by a machine that drills holes in the soil. "A child would have to eat the roots of the grass to get the poison" he is quoted as saying.

However, alert reporters soon proved these assurances to be without foundation. The actual method of application was to be by seeders, blowers and helicopters: the same type of procedure that in Illinois wiped out robins, brown thrashers and meadowlarks, killed sheep in the pastures, and contaminated the forage so that cows gave milk containing poison.

Yet at a hearing of sorts concerned Norfolk citizens were told merely that the State's Department of Agriculture was committed to the program and that it would therefore be carried out.

The fundamental wrong is the authoritarian control that has been vested in the agricultural agencies. There are, after all, many different interests involved: there are problems of water pollution, of soil pollution, of wildlife protection, of public health. Yet the matter is approached as if the agricultural interest were the supreme, or indeed the only one.

It seems to me clear that all such problems should be resolved by a conference of representatives of all the interests involved.

I wonder whether citizens would not do well to be guided by the strong hint given by the Court of Appeals reviewing the so-called DDT case of the Long Island citizens a few years ago.

This group sought an injunction to protect them from a repetition of the gypsy moth spraying. The lower court refused the injunction and the United States Court of Appeals sustained this ruling on the grounds that the spraying had already taken place and could not be enjoined. However, the court made a very significant comment that seems to have been largely overlooked. Regarding the possibility of a repetition of the Long Island spraying, the judges made this significant general comment: ". . . it would seem well to point out the advisability for a district court, faced with a claim concerning aerial spraying or any other program which may cause inconvenience and damage as widespread as this 1957 spraying appears to have caused, to inquire closely into the methods and safeguards of any proposed procedures so that incidents of the seemingly unnecessary and unfortunate nature here disclosed, may be reduced to a minimum, assuming, of course, that the government will have shown such a program to be required in the public interest."

Here the United States Court of Appeals spelled out a procedure whereby citizens may seek relief in the courts from unnecessary, unwise or carelessly executed programs. I hope it will be put to the test in as many situations as possible.

If we are ever to find our way out of the present deplorable situation, we must remain vigilant, we must continue to challenge and to question, we must insist that the burden of proof is on those who would use these chemicals to prove the procedures are safe.

Above all, we must not be deceived by the enormous stream of propaganda that is issuing from the pesticide manufacturers and from industry-related — although ostensibly independent — organizations. There is already a large volume of handouts openly sponsored by the manufacturers. There are other packets of material being issued by some of the state agricultural colleges, as well as by certain organizations whose industry connections are concealed behind a scientific front. This material is going to writers, editors, professional people, and other leaders of opinion.

It is characteristic of this material that it deals in generalities, unsupported by documentation. In its claims for safety to human beings, it ignores the fact that we are engaged in a grim experiment never before attempted. We are subjecting whole populations to exposure to chemicals which animal experiments have proved to be extremely poisonous and in many cases cumulative in their effect. These exposures now begin at or before birth. No one knows what the result will be, because we have no previous experience to guide us.

Let us hope it will not take the equivalent of another thalidomide tragedy to shock us into full awareness of the hazard. Indeed, something almost as shocking has already occurred — a few months ago we were all shocked by newspaper accounts of the tragedy of the Turkish children who have developed a horrid disease through use of an agricultural chemical. To be sure, the use was unintended. The poisoning had been continuing over a period of some seven years, unknown to most of us. What made it newsworthy in 1962 was the fact that a scientist gave a public report on it.

A disease known as toxic porphyria has turned some 5,000 Turkish children into hairy, monkey-faced beings. The skin becomes sensitive to light and is blotched and blistered. Thick hair covers much of the face and arms. The victims have also suffered severe liver damage. Several hundred such cases were noticed in 1955. Five years later, when a South African physician visited Turkey to study the disease, he found 5,000 victims. The cause was traced to seed wheat which had been treated with a chemical fungicide called hexachlorobenzene. The seed, intended for planting, had instead been ground into flour for bread by the hungry people. Recovery of the victims is slow, and indeed worse may be in store for them. Dr. W. C. Hueper, a specialist on environmental cancer, tells me there is a strong likelihood these unfortunate children may ultimately develop liver cancer.

"This could not happen here," you might easily think.

It would surprise you, then, to know that the use of poisoned seed in our own country is a matter of present concern by the Food and Drug Administration. In recent years there has been a sharp increase in the treatment of seed with chemical fungicides and insecticides of a highly poisonous nature. Two years ago an official of the Food and Drug Administration told me of that agency's fear that treated grain left over at the end of a growing season was finding its way into food channels.

Now, on last October 27, the Food and Drug Administration proposed that all treated food grain seeds be brightly colored so as to be easily distinguishable from untreated seeds or grain intended as food for human beings or livestock. The Food and Drug Administration reported: "FDA has encountered many shipments of wheat, corn, oats, rye, barley, sorghum, and alfalfa seed in which stocks of treated seed left over after the planting seasons have been mixed with grains and sent to market for food or feed use. Injury to livestock is known to have occurred.

"Numerous federal court seizure actions have been taken against lots of such mixed grains on charges they were adulterated with a poisonous substance. Criminal cases have been brought against some of the shipping firms and individuals.

"Most buyers and users of grains do not have the facilities or scientific equipment to detect the presence of small amounts of treated seed grains if the treated seed is not colored. The FDA proposal would require that all treated seed be colored in sharp contrast to the natural color of the seed, and that the color be so applied that it could not readily be removed. The buyer could then easily detect a mixture containing treated seed grain, and reject the lot."

I understood, however, that objection has been made by some segments of the industry and that this very desirable and necessary requirement may be delayed. This is a specific example of the kind of situation requiring public vigilance and public demand for correction of abuses.

The way is not made easy for those who would defend the public interest. In fact, a new obstacle has recently been created, and a new advantage has been given to those who seek to block remedial legislation. I refer to the income tax bill which becomes effective this year. The bill contains a little known provision which permits certain lobbying expenses to be considered a business expense deduction. It means, to cite a specific example, that the chemical industry may now work at bargain rates to thwart future attempts at regulation.

But what of the nonprofit organizations such as the Garden Clubs, the Audubon Societies and all other such tax-exempt groups? Under existing laws they stand to lose their tax-exempt status if they devote any "substantial" part of their activities to attempts to influence legislation. The word "substantial" needs to be defined. In practice, even an effort involving less than 5 percent of an organization's activity has been ruled sufficient to cause loss of the tax-exempt status.

What happens, then, when the public interest is pitted against large commercial interests? Those organizations wishing to plead for protection of the public interest do so under the peril of losing the tax-exempt status so necessary to their existence. The industry wishing to pursue its course without legal restraint is now actually subsidized in its efforts.

This is a situation which the Garden Club, and similar organizations, within their legal limitations, might well attempt to remedy.

There are other disturbing factors which I can only suggest. One is the growing interrelations between professional organizations and industry, and between science and industry. For example, the American Medical Association, through its newspaper, has just referred physicians to a pesticide trade association for information to help them answer patients' questions about the effects of pesticides on man. I would like to see physicians referred to authoritative scientific or medical literature — not to a trade organization whose business it is to promote the sale of pesticides.

We see scientific societies acknowledging as "sustaining associates" a dozen or more giants of a related industry. When the scientific organization speaks, whose voice do we hear — that of science or of the sustaining industry? The public assumes it is hearing the voice of science.

Another cause of concern is the increasing

size and number of industry grants to the universities. On first thought, such support of education seems desirable, but on reflection we see that this does not make for unbiased research — it does not promote a truly scientific spirit. To an increasing extent, the man who brings the largest grants to his university becomes an untouchable, with whom even the University president and trustees do not argue.

These are large problems and there is no easy solution. But the problem must be faced.

As you listen to the present controversy about pesticides, I recommend that you ask yourself — Who speaks? — And Why?

[1963]

Letter to Dr. George Crile, Jr.

Carson's cancer along with its attendant heart disease entered a more virulent phase early in 1963. This letter to her cancer surgeon and friend, George "Barney" Crile,★ of the Cleveland Clinic, was written barely a month after her triumphant appearance at the Garden Club and just shortly after the death of Crile's wife, Jane, who had been Carson's longtime friend, and who had herself fought a desperate battle against breast cancer. Carson describes her latest symptoms to Crile and courageously asks for the truth

★ George Crile, Jr., was an internationally famous surgeon who specialized in breast cancer at the Cleveland Clinic. His nickname was Barney. Carson also refers here to Kay Halle, who was Jane Halle Crile's sister, and to Dr. Caulk, who was her radiologist at the Washington Hospital Center.

from a physician for whom she had great respect and affection.

There is great irony in the fact that as she was battling the economic power and secret influence of corporate America, Carson was having to fight the medical profession to tell her the truth about her own illness. Even more tragically, Carson correctly understood the need to hide the truth of her cancer from all but a handful of friends, lest the chemical industry use her illness to discredit her scientific objectivity. In the hope of achieving a greater good, she kept silent.

Dear Barney,

You have been much in my mind and it was good to have a talk with Kay recently, to hear some of the things I wanted to know. I am glad you have the book [Crile's book on diving treasures *More Than Booty* (1965) with Jane Crile] to work on, and above all, glad you and Jane had those months to work on it together, giving it form and substance. It may be emotionally hard in some ways for you to carry it through to completion, and yet I think it will be a satisfaction.

Jane meant many things to me — a friend I loved and greatly admired, and a tower of strength in my medical problems. When she wrote me, after my visit with you two years ago, that she shared my problem, it was as though a great tide of courage flowed into me. If she, so vibrant, so gay, so full of the love of life, could live with the problem so fearlessly, I could at least try to do the same. Over the months since then the feeling I've had could best be explained by an analogy. Once, years ago, my mother and I were driving at night in uninhabited, unfamiliar country near the North Carolina coast. For the 50 or more miles through those wooded lowlands we were able to follow the lights of a car ahead. As long as it progressed smoothly I knew our way was clear. Jane was that kind of reassuring light to me. Now, without that light to follow, I admit my courage is somewhat shaken.

But you, Barney, for different reasons, are also a great source of strength. So now I'm writing you of my current problems. I didn't want to bother you while Jane was ill, and for that matter the more important ones have just happened, or at least have just been noticed.

First: I finally saw a cardiologist, Dr. Bernard Walsh, about three weeks ago. I definitely have angina (even the cardiogram is now abnormal, but he said the diagnosis was perfectly clear from symptoms alone) of the less common type in which the pains come on without physical provocation, the worst ones during sleep. Dr. Walsh said frankly the implications are serious and it is most important to get the situation under control. So — I'm virtually under house arrest, not allowed to go anywhere (except as you will see later), no stairs, no exertion of any kind. I had to rent a hospital bed for sleeping in a raised position. I'm taking peritrate. For the first ten days or so there was a big improvement, but I must admit the pains are sneaking back, though not in the night.

The second problem is in your department. About two weeks ago I noticed a tender area above the collar bone on the left (operated) side, and on exploring found several hard bodies I took to be lymph nodes. Dr. Caulk was just going out of town for several days and said he would come to the house on his return. By that time I was so sure I was going to need treatment that I just had myself taken down to see him. (This was last Wednesday.) They are definitely lymph nodes "gone bad," some lying fairly well up in the neck. This is the side opposite last year's trouble spots and is an area never previously treated. So we have begun — 5-minute treatments 3 days a week to keep my hospital trips to a minimum.

Now there is a further complication. At the time I went in about my back in December I kept making remarks about having "arthritis" in my left shoulder, but no one paid much attention. It has been increasingly painful, and now there is some difficulty about certain arm movements. I had begun to have suspicions, so now I've tackled Dr. Caulk about it again. They took a picture Friday and there does seem to be trouble. He let me see the x-rays. It is the coracoid process of the scapula — the edge of it looks irregular and sort of eroded. For some reason Dr. Caulk seems rather puzzled — says he wants some of his associates to look at it and may want a picture from another angle, but on the whole he does feel it is a metastasis.

Well, all this brings questions in my own mind, which leaps to conclusions that may or may not be justified. Oh — the back trouble cleared up, but so slowly that Dr. Caulk had about decided it wasn't a metastasis. Treatment was begun just before Christmas and completed December 31. I was still in considerable pain in mid-January. Then rather rapid improvement set in and now it's OK. But now this bone deterioration in the shoulder makes me think all the more I had a metastasis in the spine. Dr. C. says not necessarily, but I think he's just trying to reassure me.

Barney, doesn't this all mean the disease has moved into a new phase and will now move more rapidly to its conclusion? You told me last year that it might stay in the lymph nodes for years, but that if it began going into bone, etc., that would be a different story. If this is the correct interpretation I feel I need to know. I seem to have so many matters I need to arrange and tidy up, and it is easy to feel that in such matters there is plenty of time. I still believe in the old Churchillian determination to fight each battle as it comes ("We will fight on the beaches — " etc.), and I think a determination to win may well postpone the final battle. But still a certain amount of realism is indicated, too. So I need your honest appraisal of where I stand.

Jane continues to give me courage. Kay told me of her question to the doctors: "Which of you is in charge of not giving up?" How like her! Well, I nominate you to that post. [. . .]

My love to the children. As ever,

∼: 30 :∼

[1963]

The Pollution of Our Environment

Carson was invited to present the opening lecture to the Kaiser Foundation Hospitals and Permanente Medical Group in San Francisco at their annual symposium, but when the time neared for the October trip to the West Coast, Carson was debilitated from radiation treatment and frequently in pain. Nonetheless she made the trip, knowing that the symposium presented a unique opportunity to reach an influential audience.

Her official explanation for the cane she used to get on and off the stage was arthritis. The hushed and riveted audience of 1,500 physicians and health care providers did not seem to notice or care that she sat to deliver her hour-long lecture.

This was the first speech in which Carson pub-

340

licly identified herself as an ecologist. Her message emphasized the links between species and their biological and physical environment, and the dynamic systems that govern the ecosystem.

There are reverberations of Silent Spring *throughout this beautifully crafted speech, the last she gave. Carson expanded her criticism of a society that seldom evaluated the risks of new technology before it was entrenched into social systems. She also included a final warning against making the sea a dumping ground for the "poisonous garbage of the atomic age."*

[...] I suppose it is rather a new, and almost a humbling thought, and certainly one born of this atomic age, that man could be working against himself. In spite of our rather boastful talk about progress, and our pride in the gadgets of civilization, there is, I think, a growing suspicion — indeed, perhaps an uneasy certainty — that we have been sometimes a little too ingenious for our own good. In spite of the truly marvelous inventiveness of the human brain, we are beginning to wonder whether our power to change the face of nature should not have been tempered with wisdom for our own good, and with a greater sense of responsibility for the welfare of generations to come.

The subject of man's relationship to his environment is one that has been uppermost in my own thoughts for many years. Contrary to the beliefs that seem often to guide our actions, man does not live apart from the world; he lives in the midst of a complex, dynamic interplay of physical, chemical, and biological forces, and between himself and this environment there are continuing, never-ending interactions. I thought a good deal about what I could say most usefully tonight on the subject assigned to me — "The Pollution of Our Environment." Unfortunately, there is so much that could be said. I am afraid it is true that, since the beginning of time, man has been a most untidy animal. But in the earlier days this perhaps mattered less. When men were relatively few, their settlements were scattered; their industries undeveloped; but now pollution has become one of the most vital problems of our society. I don't want to spend time tonight giving a catalog of all the various kinds of pollution that today defile our land, our air, and our waters. I know that this is an informed and intelligent audience, and I am sure all these facts are known to you. Instead, I would like to present a point of view about pollution — a point of view which seems to me a useful and necessary starting point for the control of an alarming situation. Since the concept of the environment and its relation to life will underlie everything I have to say (and indeed, I think it is central to this whole symposium), I should like in the beginning to remind you of some of the early history of this planet.

I should like to speak of that strange and seemingly hostile environment that, nevertheless, gave rise to an event possibly unique in our solar system: the origin of life. Of course, our thoughts on this must be speculative; but nevertheless there is fairly wide agreement among geologists, astronomers, geochemists, and biologists about the conditions that must have prevailed just before life appeared on earth. They were, of course, very different from those of the present day. Remember, for example, that the atmosphere probably contained no oxygen; and because of that there could be no protective layer of ozone in the upper atmosphere. As a result, the full energy of the sun's ultraviolet rays must have fallen upon the sea; and there in the sea, as we know, there was an abundance of simple chemical compounds. These included carbon dioxide, methane, and ammonia, ready at hand for the complex series of combinations and syntheses that must have occurred. I shall not take time to describe the stages that presumably took place over long eons of time to produce, first, molecules capable of reproducing themselves; then some simple organisms, possibly resembling the viruses, and then doubtless much later organisms able to make their own food because of their possession of chlorophyll. Rather than stressing these details, I want to suggest two general thoughts: (1) So far as our present knowledge goes, nowhere else in the solar system have conditions equally hospitable to life occurred. This earth, then, presented an environment of extraordinary fitness; and life is a creation of that environment. (2) No sooner was life created than it began to act upon the environment. The early virus-like organisms must have rapidly reduced the supplies of nutrients adrift in that primitive ocean. But more important was the change that took place as soon as plants began the process of photosynthesis. A by-product of this process was the release of oxygen into the atmosphere. And so, gradually, over the millions and billions of years, the nature of the atmosphere has changed; and the air that we breathe today, with its rich proportion of oxygen, is a creation of life.

As soon as oxygen was introduced into the atmosphere, an ozone layer began to form, high up; this shielded the earth from the fierce energy of the ultraviolet rays, and the energy needed for the creation of new life was withdrawn.

From all this we may generalize that, since the beginning of biological time, there has been the closest possible interdependence between the physical environment and the life it sustains. The conditions on the young earth produced life; life then at once modified the conditions of the earth, so that this single extraordinary act of spontaneous generation could not be repeated. In one form or another, action and interaction between life and its surroundings have been going on ever since.

This historic fact has, I think, more than academic significance. Once we accept it we see why we cannot with impunity make repeated assaults upon the environment as we now do. The serious student of earth history knows that neither life nor the physical world that supports it exists in little isolated compartments. On the contrary, he recognizes that extraordinary unity between organisms and the environment. For this reason he knows that harmful substances released into the environment return in time to create problems for mankind.

The branch of science that deals with these

interrelations is Ecology; and it is from the viewpoint of an ecologist that I wish to consider our modern problems of pollution. To solve these problems, or even just to keep from being overwhelmed by them, we need, it is true, the services of many specialists, each concerned with some particular facet of pollution. But we also need to see the problem as a whole; to look beyond the immediate and single event of the introduction of a pollutant into the environment, and to trace the chain of events thus set into motion. We must never forget the wholeness of that relationship. We cannot think of the living organism alone; nor can we think of the physical environment as a separate entity. The two exist together, each acting on the other to form an ecological complex or an ecosystem.

There is nothing static about an ecosystem; something is always happening. Energy and materials are being received, transformed, given off. The living community maintains itself in a dynamic rather than a static balance. And yet these concepts, which sound so fundamental, are forgotten when we face the problem of disposing of the myriad wastes of our modern way of life. We behave, not like people guided by scientific knowledge, but more like the proverbial bad housekeeper who sweeps the dirt under the rug in the hope of getting it out of sight. We dump wastes of all kinds into our streams, with the object of having them carried away from our shores. We discharge the smoke and fumes of a million smokestacks and burning rubbish heaps into the atmosphere in the hope that the ocean of air is somehow vast enough to contain them. Now, even the sea has become a dumping ground, not only for assorted rubbish, but for the poisonous garbage of the atomic age. And this is done, I repeat, without recognition of the fact that introducing harmful substances into the environment is not a one-step process. It is changing the nature of the complex ecological system, and is changing it in ways that we usually do not foresee until it is too late.

This lack of foresight is one of the most serious complications, I think. I remember that Barry Commoner pointed out, in a masterful address to the Air Pollution Conference in Washington last winter, that we seldom if ever evaluate the risks associated with a new technological program before it is put into effect. We wait until the process has become embedded in a vast economic and political commitment, and then it is virtually impossible to alter it.

For example, surely it would have been possible to determine in the laboratory how detergents would behave once released into public water supplies; to foresee their nearly indestructible nature. Now, after years of use in every woman's dishwasher and washing machine, the process of converting to "soft" detergents will be a long and a costly one.

So our approach to the whole problem is shot through with fallacies. We have persisted too long in the kind of thinking that may have been appropriate in the days of the pioneers, but is so no longer — the assumption that the rivers, the atmosphere, and the sea are vast enough to contain whatever we pour into them. I remember not long ago, I heard a supposedly able scientist, the director of one of our agricultural institutions, talk glibly about the "dilution of the pollution," repeating this magical phrase as though it provided the answer to all our problems. It does not, for several reasons.

One reason, as I expect Dr. Brown will tell us tonight, is that there are entirely too many of us; and so our output of pollutants of all kinds has become prodigious. Another reason is the very dangerous nature of much of the present-day pollution. Substances that are highly capable of entering into biological reactions with living organisms. The third very important reason is that the pollutant seldom stays where we put it, and seldom remains in the form in which it was introduced.

Let us look at a few examples. The most serious problem related to modern synthetic pesticides, in my opinion, is the fact that they are becoming long-term, widespread contaminants of the environment. Some of them persist in soil for ten years or more, entering into what surely is one of the most complex and delicately balanced of all ecological systems. They have entered both surface and ground waters; they have been recovered not only from most of the major river systems but in the drinking water of many communities. Their importance as air contaminants is only beginning to be recognized. I remember this past summer there was a freak mishap in the State of Washington, which provided a rather dramatic illustration: a temperature inversion kept a very dangerous chemical, which was sprayed from the air, from settling on the crops that were being sprayed. Instead, the chemical remained in a drifting cloud for some hours and before the incident was over several cows had died of poisoning and some thirty people had been hospitalized. Then there was the incident in Long Island last winter, when several schools had to be closed because of dust from the potato fields — dust that was carrying insecticides and blowing through the screens of the school windows.

Less dramatic than those examples, but probably more important in the long run, is the fact, seldom remembered, that, for example, of all the DDT sprayed from the air less than half falls directly to the soil or to the intended target. The remainder is presumably dispersed in small crystals in the atmosphere. These minute particles are the components of what we know as "drift," or the dispersal of pesticides far beyond the point of application. This is a subject of great importance and one on which few studies have been made. We don't even know the mechanics or the mechanisms of drift. We certainly need to find out.

A few months ago, wide publicity was given to a release purporting to show that only a very small percentage of the land surface of the United States is sprayed with pesticides in any year. I don't necessarily quarrel with the statement; it may or may not be correct. But I do quarrel very seriously with the interpretation, which implies that the pesticide chemicals are confined to very limited areas; to the areas where they are applied. There are a number of reports, from many different sources, which show how inaccurate that is. The Department of the Interior, for example, has records of the occurrence of pesticide residues in waterfowl, in the eggs of the water-fowl, and in associated vegetation in far arctic regions hundreds of miles from any known spraying. The Food and Drug Administration has revealed the discovery of pesticide residues in quite substantial amounts in the liver oils of marine fishes taken far at sea, fishes of species that do not come into inshore waters. How do those things happen? We do not know. But we must remember that we are dealing with biological systems and cyclic movements of materials through the environment.

Take, for example, some of the recent demonstrations of what happens when pesticides enter a natural food chain. They progress through it in a fashion that is really explosive. You have several examples here in the State of California, at the Tule Lake and Klamath National Wildlife Refuges. Water entering the refuges from surrounding farms is carrying in residues of insecticides. These have now become concentrated in food chain organisms and in recent years have resulted in a heavy mortality among fish-eating birds.

Then, at Big Bear Lake in San Bernardino County, toxophene was applied to the lake at a concentration of only 0.2 of 1 part per million. But notice how it was built up. Four months later it was concentrated in plankton organisms at a level of 73 parts per million. Later, residues in fish reached 200 parts per million. In a fish-eating bird, a pelican, 1700 parts per million.

And at Clear Lake, not far from here, efforts to control the gnat population have had a long and a troubled history. Beginning in 1949, the chemical DDD was applied to the lake in very low concentrations. It was later picked up by the plankton, by plankton-eating fish, and by fish-eating birds. The maximum application to the water itself was only $\frac{1}{50}$ part per million; yet in some of the fishes the concentration reached 2500 parts per million. The western grebes which nested on the shore of the lake and are fish-eaters almost died out. When their tissues were analyzed they were found to contain heavy concentrations of the chemical. A very interesting phenomenon was that five years after the last application of the chemical, although the water of the lake itself was free of the poison, the chemical apparently had gone into the living fabric of the lake; all of the resident plants and animals still carried the residues and were passing them on from generation to generation.

One of the most troublesome of modern pollution problems is the disposal of radioactive wastes at sea. By its very vastness and seeming remoteness the sea has attracted the attention of those faced with the problem of disposing of the by-products of atomic fission. And so the ocean has become a natural burying-place for contaminated rubbish and for other low-level wastes of the atomic age. Studies to determine the limits of safety in this procedure for the most part have come after rather than before the fact, and disposal activities have far outrun our precise knowledge as to the fate of these waste products.

If disposal of radioactive wastes at sea is to be safe, the material must remain approximately where it is put, or else it must follow predictable paths of distribution, at least until the decay of the radioactive substances has reduced them to relatively harmless levels. The more we know about the depths of the sea, the less do they appear to be a place of calm where deposits may remain undisturbed for centuries. There is far greater activity at deep levels than we formerly suspected. Below the known and charted surface currents there are others which run at their own speeds, in their own directions, and with their own volume. There are powerful turbidity currents that rush down over the continental margins. Even on the ocean floor, at great depths, moving waters are constantly sorting over the sediments, leaving the evidence of their work in ripple marks.

All of these activities, plus the long recognized upwelling of water from the depths and the opposite, downward sinking of great masses of surface water result in a gigantic mixing process. When we dump radioactive wastes in the sea we are introducing them into a dynamic system. But this transport by the sea is only part of the problem, because marine organisms also play an important part in concentrating and distributing radioisotopes. We still need to learn a great deal about the processes involved when radioactive materials are introduced, through fallout, into the marine environment. The studies that have been made reveal movements of great complexity between sea water and the hordes of plankton creatures, between the plankton and the organisms higher in the food chain, between the sea and the land and from the land to the sea.

The most important fact about this is that the marine organisms bring about a marked distribution, both vertical and horizontal, of the radioactive contaminants. As the plankton make regular migrations, sinking into deep water in the daytime and rising to the surface at night, with the organisms go the radioisotopes they have absorbed, or that may adhere to them. As a result, the contaminants are made available to other organisms in new areas; and as they are taken up by larger, more active animals, they are subject to transport over long horizontal distances; migrating fishes, seals, and whales may distribute radioactive materials far beyond the point of origin.

All these facts have important meaning for us. They show that the contaminant does not remain in the place deposited, or in its original concentration, but rather becomes involved in biological activities of an intensive nature.

It is surprising, then, that so little thought seems to have been given to the biological cycling of materials in one of the most crucial problems of our time: the understanding of the true hazards of radiation and fallout. There have been situations in the news in recent months that are perfect illustrations of our lack of application of the ecological understanding that we have. I think one of the best examples of what I mean is taking place now in the arctic regions in both eastern and western hemispheres. Only two or three years ago it was reported that both the Alaskan Eskimos and the Scandinavian Lapps are carrying heavy burdens of both Sr^{90} and Ce^{137}. This is not because fallout is especially heavy in these far northern regions; indeed, it is lighter there than in areas of heavier rainfall somewhat farther south. The reason is that these native peoples occupy a terminal position in a unique food chain. This begins with the lichens of the arctic tundras; it continues through the bones and the flesh of the caribou and the reindeer, and at last ends in the bodies of the natives, who depend heavily on these animals for meat. Because the so-called "reindeer moss" and other lichens receive nutrients directly from the air, they pick up large amounts of the radioactive debris of fallout. Lichens, for example, have been found to contain 4 to 18 times as much Sr^{90} as sedges, and 15 to 66 times the Sr^{90} content of willow leaves. They are long-lived, slow-growing plants; so they retain and they concentrate what they take out.

Cesium^{137} also travels through this arctic food chain, to build up high values in human bodies. As you remember, cesium has about the same physical half-life as Sr^{90}, although its stay in the human body is relatively short, only about 17 days. However, its radiation does take the form of the highly penetrating gamma rays, thus making it potentially a hazard to the genes. About 1960 it was reported that Norwegians and also the Finnish and Swedish Lapps were carrying heavy body burdens of Ce^{137}. Then, during the summer of 1962, a team from the

Hanford Laboratories in Washington went up into the arctic and measured the levels of radioactivity in about 700 natives in 4 different villages above the arctic circle. They found that the averages for Ce^{137} were about 3 to 80 times the burden in individuals who had been tested at Hanford. In one little village, where caribou is a major item of diet, the average burden of Ce^{137} was 421 nanocuries;* the maximum burden was 790. The counts for 1963, which extended over a wider geographic area in Alaska, are said to have been still higher.

This situation almost certainly existed from the beginning of the bomb tests; yet somehow it does not seem to have been anticipated, or at least it was not widely discussed and acted upon, though the Scandinavian countries have been rather active in their investigations.

Another example which has become familiar to many of us in recent months is provided by radioactive iodine. This must always have been an important constituent of fallout, so we wonder why its significance has been largely ignored until very recently. Probably the answer lies in its very short half-life, which is only about 8 days, and in the assumption

* One billionth (10^{-9}) of a curie, a unit of radioactivity.

that decay would have rendered it harmless before it could affect human beings. But the facts, of course, are otherwise. Radioactive iodine is a component of the lower atmospheric fallout and so, depending upon weather conditions it may reach the earth so early that much of its radioactivity is retained. Its distribution may be spotty, also, because of wind, rain, or other weather conditions. So we have the occurrence of the so-called "hot spots."

But we are not primarily concerned with the amount of radioactive iodine on the ground. It is not believed that we absorb significant amounts through the skin or even by inhalation. What is important is the entrance of this material into the food chains. From that point the route to the human body is short and direct. From contaminated pasture grass to the cow, from fresh cow's milk to the human consumer; and once in the body, the iodine finds its natural target, the thyroid gland. It follows that small children, with their small thyroids and their relatively large intake of milk, are endangered more than are adults.

Only a few years ago, it was declared by a scientist testifying before the Committee on Atomic Energy that radioiodine from worldwide fallout is not a problem of concern to humans, and it is not expected that it will become a problem in the future. At the time this prediction was made, there was no national system for sampling. Most of the sampling done since then seems to have suffered from various defects. For example, data on milk supplies for large cities have little meaning, because such milk is a mixture of collections from various areas and the occasional high levels of contamination are easily obscured. Until the summer of 1962 no attempt seems to have been made to collect fallout data and milk contamination data at the same place and time. It appears also that much of the monitoring data reported by the AEC refers to measurements of gamma-ray intensity from the ground, or of beta radioactivity near the ground or in the air. However, as we have seen, what is important is not the radioactive source outside the body, but the entrance of the radioactivity into the food chain and so into the human body.

In the summer of 1962, the Utah State Department of Health began to make its own evaluation of this problem and quickly decided that a hazardous situation existed. All of the five bomb tests carried out in Nevada in July 1962 had carried radioactive iodine into Utah. As exposures began to exceed the yearly radiation protection guide, the state recommended protective measures. Of course, for radioactive iodine, these are very simple: cows may be transferred to stored feed; contaminated milk may be diverted to processing plants for use in ways that will assure an appropriate lapse of time before it is consumed. Knapp, of the AEC's Division of Biology and Medicine, made other observations in Utah, examining single samples of milk rather than dealing in averages, or in composite samples. And these studies bore out the contention that high levels of radioactive iodine were occurring in certain areas. The Utah situation is probably not unique. A few months ago the Committee for Nuclear Information testified before the Joint Committee on Atomic Energy and declared that a number of local populations, especially in Nevada, Utah, Idaho, and probably other communities scattered throughout the continental United States, have been exposed to fallout of medically unacceptable proportions, especially in the cases of children who drink fresh, locally produced milk. The evidence provided by the Committee, as well as that collected in the recently released Knapp report, would seem to support this conclusion. Yet as recently as May of this year the Public Health Service stated that I^{131} doses from weapons testing have not caused undue risk to health.

My reason for reviewing these facts, which I am sure are familiar to most of you, is simply to emphasize that we have not yet become sophisticated enough to view these matters as the ecological problems [. . .] they are. Of course there are various ways of studying the problem; there are various angles from which it must be approached, and what I am suggesting does not necessarily preclude other approaches. But I think that the ecological aspect of it must be considered. We must remember that we have introduced these things into dynamic systems that comprise our environment, and it is not enough to monitor the entrance of the contaminant into the environment at that single point. We must be prepared, with the best understanding of all concerned — the physician, the biologist, the ecologist — to follow the contaminants through whatever path they take, through physical and biological systems. This demands more extensive studies than any that have been undertaken, more comprehensive monitoring programs, and more realistic evaluations.

In my opinion, we have been too unwilling to concede the possibility of hazard or of the actual existence of hazard. We have been too unwilling to give attention to the preparation of countermeasures to cope with the hazardous situations when they do arise. Perhaps not now, but perhaps in the future that will arise. Indeed, a report to the Surgeon General by the National Advisory Committee on Radiation in 1962 revealed that except in the case of I^{131}, no effective countermeasures exist. In the climate of euphoria that is generated by repeated assurances that all is well, there is little public support and there is little money for the kind of research that needs to be done. I, for one, would like to see the public considered [. . .] capable of hearing the facts about the hazards that exist in the modern environment. I should like to have them considered capable of making intelligent decisions as to prudent and necessary measures that ought to be taken.

Currently, in this specific area of radiation hazards, I think there is a certain danger that we will feel that the recent test ban treaty makes the whole fallout problem obsolete. This, in my opinion, is not true. The longer-lived isotopes will remain in the upper atmosphere for years to come, and we are still destined to receive heavy fallout from past tests. Another very important point is that underground tests have been known to produce atmospheric contamination through venting in the past, and they will almost certainly continue to do so.

The third point is that environmental contamination by radioactive materials is apparently an inevitable part of the atomic age. It is an accompaniment of the so-called "peaceful" uses of the atom as well as of the testing of weapons. This contamination will come about occasionally by accidents, and perpetually by the disposal of wastes.

Underlying all of these problems of introducing contamination into our world is the question of moral responsibility — responsibility, not only to our own generation, but to those of the future. We are properly concerned about somatic damage to generations now alive; but the threat is infinitely greater to the generations unborn; to those who have no voice in the decisions of today, and that fact alone makes our responsibility a heavy one.

I recently read some calculations made by Professor H. J. Muller. His general conclusion was that the amount of somatic damage from radiation as it is distributed today is far less than the damage which this same radiation, received and transmitted by the present generation, will inflict upon posterity. His further conclusion was that hereditary damage should be the chief touchstone in the setting of permissible or acceptable dose limits. But apparently we have a long way to go and much enlightenment to gain before any agreement can be reached on standards of this kind.

The question of genetic damage from harmful elements in the environment is one that particularly interests me. Elsewhere I have made the suggestion that pesticide chemicals should be viewed with great suspicion as possible agents of genetic damage to man. This suggestion has been challenged by some on the grounds that there is no proof that these chemicals are having such an effect. I don't believe we should wait for some dramatic demonstration before making a thorough study of the potential genetic effect of all chemicals that are widely introduced into the human environment. By the time such a discovery is made otherwise, it will be too late to eradicate them. Some of the chemicals that are now in use as herbicides and insecticides do have mutagenic effects on lower organisms. Others have the ability to cause chromosome damage or a change in chromosome number, and as you know this type of chromosomal abnormality may be associated with a wide variety of congenital defects in man, including mental retardation. I think we should test the pesticide chemicals on several of the organisms that reproduce rapidly and so lend themselves to genetic experiments. If the chemicals then prove mutagenic, or otherwise disruptive of genetic systems in a variety of test organisms, then I think we should withdraw them from use. I am not impressed with the argument that they might not have similar effects in man. After all, the science of genetics was founded when an obscure Austrian monk performed some experiments on garden peas; and the basic hereditary laws he discovered have proved generally applicable in both plants and animals.

Again, another fact of far-reaching significance, that influences in the external environment can cause mutations, was discovered by Professor Muller in experiments on an insect; yet few doubt its applicability to man. Indeed, one of the most striking phenomena in biology is the basic similarity of genetic systems throughout the living world. Yet again and again, in this whole field of environmental influences in relation to life, and this includes our theme of pollution and its impact on life, we meet a strange reluctance to concede that man is, himself, susceptible to harm. It may be admitted freely, for example, that an agricultural chemical entering a river could kill thousands of fish; but it will be denied that this chemical could do any harm to the person who might drink the water.

Reports of the decimation of whole populations of birds are shrugged off with the
thought that it can't happen to us. If we carry
this view to its logical conclusion, it would
make a mockery of all the elaborate testing,
involving millions of laboratory animals; yet I
have been astonished to discover how frequently it crops up, if not stated directly, then
at least as the implied basis for an official
point of view or decision, or perhaps more
often for the lack of any decisive action. I
wonder sometimes whether this attitude may
not have a deep significance which is relative
to our theme tonight. It seems to me to imply
a sort of rejection of our past — a reluctance
or an unreadiness to accept the fact that man,
like all other living creatures, is part of the vast
ecosystems of the earth, subject to the forces
of the environment.

As I look back through history I find a parallel. I ask you to recall the uproar that followed Charles Darwin's announcement of his
theories of evolution. The concept of man's
origin from pre-existing forms was hotly and
emotionally denied, and the denials came not
only from the lay public but from Darwin's
peers in science. Only after many years did
the concepts set forth in *The Origin of Species*
become firmly established. Today, it would
be hard to find any person of education who

would deny the facts of evolution. Yet so many of us deny the obvious corollary: that man is affected by the same environmental influences that control the lives of all the many thousands of other species to which he is related by evolutionary ties.

I find it quite fascinating to speculate what hidden fears in man, what long-forgotten experiences, have made him so loath to acknowledge first, his origins and then his relationship to that environment in which all living things evolved and coexist. The Victorians at last freed themselves from the fears and superstitions that made them recoil in shock and dismay from Darwinian concepts. And I look forward to a day when we, also, can accept the facts of our true relationship to our environment. I believe that only in that atmosphere of intellectual freedom can we solve the problems before us now.

Thank you.

∿ 31 ∿

[1963]

Letter to Dorothy Freeman

Carson spent what would be her final summer in Maine, hoping for more time to say all the things she wanted to say, but knowing it was an ephemeral hope. She planned a book on evolutionary biology, but most of all she wanted time to expand her 1956 article "Help Your Child to Wonder" into a book on the value and necessity of a sense of wonder in the modern world.

Although she rarely spoke of her illness, she was able to write about death through her understanding of the rhythms, enduring cycles, and patterns of the natural world.

This letter, written to her friend Dorothy Freeman, after the two had spent a sunlit morning at Newagen, one of their favorite places along the shore of the Sheepscot, was intended to acknowledge her approaching death and to comfort her friend. With Freeman's permission it was read by the Reverend Duncan Howlett at the memorial service Carson asked him to hold after her death. Rachel Carson died at her home in Maryland of cancer and heart disease on April 14, 1964, at the age of fifty-six.

Dear One,

This is a postscript to our morning at Newagen, something I think I can write better than say. For me it was one of the loveliest of the summer's hours, and all the details will remain in my memory: that blue September sky, the sounds of wind in the spruces and surf on the rocks, the gulls busy with their foraging, alighting with deliberate grace, the distant views of Griffiths Head and Todd Point, today so clearly etched, though once half seen in swirling fog. But most of all I shall remember the Monarchs, that unhurried westward drift of one small winged form after another, each drawn by some invisible force. We talked a little about their migration, their life history. Did they return? We thought not; for most, at least, this was the closing journey of their lives.

But it occurred to me this afternoon,

remembering, that it had been a happy spectacle, that we had felt no sadness when we spoke of the fact that there would be no return. And rightly — for when any living thing has come to the end of its life cycle we accept that end as natural.

For the Monarch, that cycle is measured in a known span of months. For ourselves, the measure is something else, the span of which we cannot know. But the thought is the same: when that intangible cycle has run its course it is a natural and not unhappy thing that a life comes to its end.

That is what those brightly fluttering bits of life taught me this morning. I found a deep happiness in it — so, I hope, may you. Thank you for this morning.

Credits

Part One

1. "Undersea" from *The Atlantic Monthly*, vol. 160 (September 1937), pp. 322–325.

2. "My Favorite Recreation" from *St. Nicholas Magazine*, vol. 49 (July 1922), p. 999.

3. "Fight for Wildlife Pushes Ahead" from the *Richmond Times-Dispatch Sunday Magazine*, March 20, 1938.

"Chesapeake Eels Seek the Sargasso Sea" from the *Baltimore Sunday Sun*, October 9, 1938.

4. Ace of Nature's Aviators. (1944) Manuscript. Rachel Carson Papers, Yale Collection of American Literature, Beinecke Rare Book and Manuscript Library, Yale University, New Haven, Connecticut. [Hereafter cited as Rachel Carson Papers.]

5. Road of the Hawks. (1945) Unpublished fragment. Rachel Carson Papers.

6. An Island I Remember. (1946) Unpublished fragment. Rachel Carson Papers.

7. "Mattamuskeet: A National Wildlife Refuge" from *Conservation in Action*, no. 4. U.S. Fish and Wildlife Service. Washington, D.C.: U.S. Government Printing Office,

1947. Illustrated by Katherine L. Howe.

Part Two

8. Memo to Mrs. Eales on *Under the Sea-Wind*. (ca. 1942) Rachel Carson Papers.

9. "Lost Worlds: The Challenge of the Islands" from *The Wood Thrush*, vol. 4, no. 5 (May–June 1949), pp. 179–187.

10. Speech given at the *New York Herald-Tribune* Book and Author Luncheon, October 16, 1951, New York, New York. Rachel Carson Papers.

11. Jacket notes for the RCA Victor recording of Claude Debussy's *La Mer* by the NBC Symphony Orchestra, Arturo Toscanini, conductor, 1951. Rachel Carson Papers.

Speech given at the National Symphony Orchestra Benefit Luncheon, September 25, 1951, Washington, D.C. Rachel Carson Papers.

12. Remarks at the acceptance of the National Book Award for nonfiction, January 29, 1952, New York, New York. Rachel Carson Papers.

13. Design for Nature Writing. Remarks made on acceptance of the John Burroughs Medal for excellence in nature writing, April 7, 1952, New York, New York, from *The*

Atlantic Naturalist (May–August 1952), pp. 232–234.

14. "Mr. Day's Dismissal" from *The Washington Post*, April 22, 1953, p. A26.

15. Preface to the Second Edition of *The Sea Around Us*, by Rachel Carson (New York: Oxford University Press, 1961).

Part Three

16. "Our Ever-Changing Shore" from *Holiday*, vol. 24 (July 1958), pp. 71–120.

17. Four Fragments from Carson's Field Notebooks. (1950–1952). Rachel Carson Papers.

18. The Edge of the Sea. Paper presented to the American Association for the Advancement of Science Symposium, "The Sea Frontier," December 29, 1953, Boston, Massachusetts. Rachel Carson Papers.

19. The Real World Around Us. Speech given at the Theta Sigma Phi Matrix Table Dinner, April 21, 1954, Columbus, Ohio. Rachel Carson Papers.

20. "Biological Sciences" from *Good Reading* (New York: New American Library, 1956).

21. Two Letters to Dorothy and Stanley Freeman, August 8, 1956, and October 7, 1956, from *Always Rachel: The Letters of*

Rachel Carson and Dorothy Freeman, edited by Martha Freeman (Boston: Beacon Press, 1995).

22. The Lost Woods. A Letter to Curtis and Nellie Lee Bok, December 12, 1956. Rachel Carson Papers.

23. Clouds. Script written for the Ford Foundation's TV-Radio Workshop, "Something About the Sky," CBS *Omnibus*, March 11, 1957. Rachel Carson Papers.

24. "Vanishing Americans" from *The Washington Post*, April 10, 1959, p. A26.

25. "To Understand Biology" from *Humane Biology Projects* (New York: The Animal Welfare Institute, 1960).

Preface to *Animal Machines: The New Factory Farming Industry*, by Ruth Harrison (London: Vincent Stuart, LTC., 1964).

26. "A Fable for Tomorrow" from *Silent Spring*, by Rachel Carson (Boston: Houghton Mifflin Co., 1962), pp. 1–3.

27. Speech given to the Women's National Press Club, December 5, 1962, Washington, D.C. Rachel Carson Papers.

28. A New Chapter to *Silent Spring*. Speech given to the Garden Club of America, January 8, 1963, New York, New York. Published in the *Bulletin of the Garden Club of America* (May 1963). Rachel Carson Papers.

29. Letter to Dr. George Crile, Jr., February 17, 1963. Rachel Carson Papers.

30. The Pollution of Our Environment. Speech given to the Kaiser-Permanente Symposium, "Man Against Himself," October 18, 1963, San Francisco, California. Rachel Carson Papers.

31. Letter to Dorothy Freeman, September 10, 1963, from *Always Rachel: The Letters of Rachel Carson and Dorothy Freeman*, edited by Martha Freeman (Boston: Beacon Press, 1995).

Acknowledgments

In the course of writing Rachel Carson's life I discovered the varied tapestry of her formal and informal writing that for one reason or another was lost to the archives or which had once been published but was now long forgotten and out of print. From the freshness of her early nature writing to the richness of her speeches as a mature literary figure, the body of this writing impressed me and made me think that others might find in these, as in her other published writing, much to treasure.

Happily Deanne Urmy, Executive Editor of Beacon Press, shared my enthusiasm for this unknown collection and lent her own deep interest in the subject so that this anthology could become reality. She has enriched the always difficult process of selection by her perceptive editorial eye and her literary discernment for which I am deeply grateful. Working with her has been a gift.

My literary agent and the trustee of Carson's literary estate, Frances Collin responded to this project with appreciation and insight, and willingly lent her invaluable archives. To Marsha S. Kear, administrative assistant to Collin, I owe a debt of several years' standing for finding obscure letters and accurate data whenever I came up empty handed.

As in my earlier work on Carson, Paul Brooks has once again graciously and with unfailing literary taste led the way. Several of Carson's unpublished excerpts first appeared in whole or in part in his splendid literary biography *The House of Life: Rachel Carson at Work*. I have chosen to republish them here because of the quality of Carson's writing they exhibit, and the insight they give on her development as a natural scientist.

Almost all of the writing I have selected first came to my notice during the years I spent working on the Papers of Rachel Carson at the Beinecke Rare Book and Manuscript Library at Yale University. I continue to owe the curators and archivists there a debt of gratitude for their knowledgeable assistance.

Several selections required scientific annotation in order to bring currency to Carson's original research. I am indebted to the following scientists who helped me verify material and introduced me to the latest thinking on scientific issues that remain controversial: David G. Smith, Department of Vertebrate Zoology, and Christopher Milensky, Division of Birds, National Museum of Natural History, Smithsonian Institution; Richard H. Backus and William Watkins, Woods Hole Oceanographic Institution, and especially George M. Woodwell, Woods Hole Research Center, who patiently responded to my inquiries and always knew where to send me; Tom Cochran, National Resources Defense Council; Cliff Curtis, World Wildlife Federation; and Matthew Perry, Patuxent Wildlife Research Center, who continually expands my understanding of wildlife management.

I was able to work on this project with the assistance and collegiality of Pamela Henson, Office of Smithsonian Institution Archives, and in the good company of the staff of the Joseph Henry Papers, Office of Smithsonian Institution Archives, to whom I continue to owe many happy debts.

My two Ruths, Ruth Brinkmann Jerome and Ruth Jury Scott, have graced my life in different ways but with infinite richness. Ruth Scott was one of Carson's compatriots but she is also a mentor and guide without whose support my world and Rachel Carson's would never have coincided so seamlessly or so happily. Ruth Brinkmann Jerome, my dear friend of forty years, began nurturing me as a young undergraduate. She remains my guide to how to live one's life with grace, humor, courage, and faithfulness.

My husband, John W. Nickum, Jr., to whom this book is dedicated, knows the richness of the support he has given to me day by day and year by year so that I could have the freedom to create and the discipline to persevere. I hope he knows some measure of my love and gratitude as well.

The employees of Thorndike Press hope you have enjoyed this Large Print book. All our Large Print titles are designed for easy reading, and all our books are made to last. Other Thorndike Press Large Print books are available at your library, through selected bookstores, or directly from the publishers.

For more information about titles, please call:

(800) 257-5157

To share your comments, please write:

Publisher
Thorndike Press
P.O. Box 159
Thorndike, Maine 04986